DR. JOE'S
BRAIN SPARKS

I79 Inspiring and Enlightening Inquiries
Into the Science of Everyday Life

JOE SCHWARCZ, PhD

Director, McGill University
Office for Science and Society

ANCHOR CANADA

Library and Archives of Canada Cataloguing in Publication
has been applied for.

ISBN: 978-0-385-66932-0

Text design: Leah Springate
Cover design: Paul Dotey
Cover images: (sparks) © Josh Westrich/Corbis;
(brain) © Cammeraydave/Dreamstime.com
Printed and bound in the USA

Published in Canada by Anchor Canada,
a division of Random House of Canada Limited

Visit Random House of Canada Limited's website: www.randomhouse.ca

10 9 8 7 6 5 4 3 2 1

CONTENTS

INTRODUCTION

For my thirteenth birthday, I received a typewriter as a present. It wasn't even electric, but at the time it was quite a gift. Today, I suspect that most students wouldn't even recognize a typewriter if they saw one. And why should they? That dinosaur has been completely replaced by the computer. But my first typewriter was a treasure (in no small part because my handwriting was atrocious—which it still is). I enjoyed writing even then, but sometimes I couldn't even read what I had written. My new typewriter changed that in a hurry. I could now write, and people could actually read what I had written.

So what sort of stuff was I interested in writing in those days? Well, call me a nerd, but I enjoyed making up science quizzes for my friends. Not exactly a common hobby for a kid in grade seven, especially given that it had nothing to do with any school activity. Science there was nonexistent. Until a general course on the subject in grade nine, I don't think I heard the word "science" mentioned. But by that time I had already become enthralled.

I had discovered science sometime in grade six, despite my teachers' best efforts to hide the existence of this discipline. It was a teenage magician at a birthday party—and not a very good one, as I recall—who involuntarily sparked my interest by pretending to fuse three pieces of rope into a single strand using a "magic chemical." To this day I do not know why he used the term "magic chemical" instead of "hocus pocus" or "abracadabra," but I do know that it set me wondering about both magic and chemistry. I took out books from the library and started to read voraciously about both.

For me, chemistry turned out to be magical. I was immediately taken by its fascinating history and its applications to daily life. A whole new world of paints, plastics, drugs, explosives, cosmetics and nutrients opened up for me. But it was somewhat of a lonely world since none of my friends were into chemistry. So I started to make up little quizzes to see if I could generate some interest. If only they realized how fascinating those cavorting molecules were, they would surely share in my excitement. My little handwritten attempts, I must admit, didn't meet with much success. And then along came the typewriter. I wouldn't say I was besieged to produce more and more of my quizzes, but at least my friends were reading them. Probably because they now *could* read them.

Truthfully, I can't remember all the questions I concocted on that first typewritten quiz of mine—it was just too long ago. I do, though, remember the very first one. That's because it was about a quote that had really struck me, and one to which I have referred on many occasions since.

Here was the question: What is the "it" in the following quotation? "It is enough to drive one mad. It gives me the impression of a primeval tropical forest full of the most remarkable things, a monstrous and boundless thicket with no way of escape, and into which one may well dread to enter."

Of course, none of my friends knew the answer to that, and how would they? "It" was organic chemistry. The quote is attributed to the German chemist Friedrich Wöhler and appeared in a letter to his former professor and mentor, the Swedish chemist Jacob Berzelius. I thought it was a great quote; I could really relate to it. I had stumbled upon organic chemistry in my readings and was beginning to realize how fascinating it was. But did it ever seem complicated! I didn't know what to make of all those drawings that somehow depicted molecules. Wöhler, I thought, had hit the nail right on the head. And I thought his picturesque words would interest my friends.

I could hardly wait to elaborate on the answer. Why? Because I was ready to say that, while Wöhler had suggested there was no escape from the thicket of organic chemistry, he himself would eventually be the one to find it. Wöhler's classic experiment in 1828, whereby he produced urea, an organic compound, in the laboratory, showed that organic molecules had no "vital force" infused in them. This dispelled the notion held by many, including Berzelius, that organic compounds could only be created by nature, not by man. The field was now thrown open. Dyes, medicines and who knew what else could be created in the lab by chemists, if they were ingenious enough. Alas, my friends didn't get too excited about it. But I remain as enthralled by organic chemistry today as I was then. More than ever, I enjoy making up questions that I hope stimulate some scientific interest. And this book is filled with them.

So, let the brain sparks fly!

CHEMISTRY: THE BEST MEDICINE

A drug introduced in 1981, isolated from a soil fungus, revolutionized the transplantation of human organs. What drug was that?

Cyclosporin. Early transplants were plagued by organ rejection. The recipient's immune system considered the organ a foreign intruder and mobilized its forces for battle. Doctors realized that if transplants were ever to be successful, the body's immune system would have to be held in check.

By the time Dr. Christiaan Barnard performed the world's first heart transplant in 1967, drugs that curbed immune activity were available, but they left a lot to be desired. The medications managed to keep the immune system from rejecting the organ, but the extent of immune suppression was such that it left the patient susceptible to all sorts of infections. In fact, Louis Washkansky, the first recipient of a transplanted heart, died of pneumonia he contracted because of suppressed immunity. But the problem of rejection was essentially solved when cyclosporin came onto the scene in 1981.

The discovery of the first truly effective anti-rejection drug was somewhat serendipitous, and dates back to the early 1970s. In those days, pharmaceutical companies searched high and low for novel antibiotics, investigating whatever fungus they could get their hands on. After all, antibiotics isolated from fungi, such as penicillin and streptomycin, had already proven their worth. Hoping to find some novel antibiotic-producing fungus, pharmaceutical companies routinely asked their employees to bring back soil samples from their travels. The Sandoz company lucked out. A soil sample collected in Norway yielded a strain of fungus that produced a compound composed of a ring of amino acids, eventually named cyclosporine. It looked like a good candidate for antibiotic activity. Unfortunately, it turned out not to have any such properties.

Eventually, though, disappointment turned to elation when cyclosporine was found to have a marked immunosuppressive effect! Administering the drug presented a problem, however, since it was almost completely insoluble in water. When taken orally, it never made it into the bloodstream. But researchers discovered that dissolving the drug in olive oil did the trick. In 1978, the first kidney and bone marrow transplants in which cyclosporine successfully prevented rejection were performed in England. Today, hearts, kidneys, livers and bone marrow are routinely transplanted, thanks in large part to cyclosporine.

There is a footnote to the cyclosporine story, and a rather significant one. After a transplant, monitoring the use of all drugs taken by patients is critical because some medications can interfere with the action of cyclosporine. St. John's wort, for example, an herbal remedy available without a prescription, can negate the effect of cyclosporine and result in rejection. This interaction came to light when a heart transplant patient's body rejected the donated organ even though an appropriate amount of cyclosporine had been administered. Unknown to his physicians, he had been taking St. John's wort purchased at a health food store to ward off his depression. He almost warded off his new heart.

Around 1000 BC, a Chinese monk introduced the idea of blowing a substance up the nose of people to protect them from smallpox. What was this substance?

A powder made from the scabs of pustules on the skin of people who had survived smallpox. The eradication of this horrific disease, which is thought to have first appeared around 10,000 BC, is one of the greatest triumphs of medicine. The last recorded case of smallpox occurred in Somalia in 1977, more than thirty years ago. How did this triumph come about? Simple: vaccination! The name associated with the introduction of the smallpox vaccine is Dr. Edward Jenner, an English country physician who acted on the observation that milkmaids who had come down with a disease known as cowpox never contracted smallpox. Jenner injected young James Phipps with material taken from a milkmaid's cowpox pustule and then exposed him to smallpox. (Obviously, there were no ethics committees at the time to approve research.) The boy didn't get the disease, and the era of vaccination, the term deriving from the Latin for *cow*, was on its way. Although Jenner usually gets the credit for introducing the smallpox vaccine, it was twenty years earlier that Benjamin Jesty, a farmer, inoculated his wife with the cowpox virus and showed that it protected her from the disease. Unfortunately, he didn't have enough oomph to influence the medical community.

Even more amazing is that a technique known as variolation had been introduced by a Chinese monk almost two thousand years earlier. After the death of the son of a high-ranking Chinese official, the monk sought a way to cure the scourge of smallpox. He hit upon the idea of blowing the dust made from ground-up pustules taken from the skin of smallpox victims up the nose of healthy people. In all likelihood this was predicated on the observation that people who had survived smallpox became immune to the disease. Lady Wortley Montague learned of this technique when her husband had a political posting in Turkey. She brought it to the attention of the British royal family and suggested that variolation could be tested on condemned prisoners. Indeed, four such men were treated, and months later were exposed to smallpox. All four survived. This was enough to convince the royal family to undergo variolation. The French thought the English were crazy. In fact, Voltaire opined that "the English are fools, they give their children smallpox to prevent their catching it." They weren't fools. In smallpox survivors the virus becomes weakened and can offer protection to others, with only a small risk of causing the actual disease. The death rate from smallpox was usually somewhere between 20 per cent and 40 per cent, but the death rate from variolation was only about I per cent. It is interesting to note that this ancient technique saved many from contracting smallpox long before it was replaced by Jenner's more effective vaccination.

"It may be that the world's oldest medicine is the earth itself." To what does that statement refer?

The ingestion of clay to absorb toxic substances. Terra sigillata, which literally means "earth that has been stamped with a seal," was originally dug up each year only on August 6, on the Greek island of Lemnos. It was mixed with the blood of a sacrificial goat, shaped into lozenges and dried. The famous Greek physician Galen recommended it as an antidote to poisons way back in the second century AD. Kings and popes commonly ate terra sigillata with their meals. Clays really do have the ability to bind various substances, and they even exhibit a property known as "cation exchange." This means they can absorb positively charged ions such as those of mercury and lead, which are highly toxic. Indeed, terra sigillata used to be an antidote to poisoning by mercuric chloride. A story from the sixteenth century speaks of a condemned German criminal who, in a bid to avoid execution, proposed an experiment to the court. He would act as a human guinea pig and take a potentially lethal amount of mercuric chloride, followed by terra sigillata in wine. If he survived, he would be released. Although he went through a torturous experience, the man did survive and was indeed freed! Today, refined clays, as in Kaopectate, are used to treat diarrhea caused by bacterial toxins in the gut. The clay can absorb the toxins and relieve the condition.

Both George Bush Sr. and his wife, Barbara, were diagnosed with Graves' disease. This initiated the testing of the water in the White House for what substance?

Iodine. Graves' disease is a form of hyperthyroidism, a condition in which the thyroid gland produces too much thyroid hormone. As a result, metabolism speeds up, leading to weight loss, insomnia, muscle weakness, tremors, sweating, frequent loose stools, palpitations, bulging eyes and a feeling of edginess. Thyroid hormone contains iodine, and it's conceivable that too high an intake of iodine can cause it to be overproduced. But the usual cause of Graves' is an autoimmune reaction in which antibodies in the blood stimulate the overproduction of thyroid hormones. Since the odds of a husband and wife both developing this type of autoimmune reaction at roughly the same time are about three million to one, the White House water was tested for iodine. No excess was found, so it seems that George and Barbara were that unlucky couple in three million who simultaneously came down with Graves' disease due to an autoimmune reaction. Other possibilities, such as a pituitary tumour, were ruled out. Autoimmune reactions can be triggered by a viral or bacterial infection, so the first couple may have shared some infectious agent.

The president and first lady received plenty of good-natured advice from the public about what to do, including eating broccoli, which is known to contain goitrogens, compounds that interfere with thyroid function. While this makes some theoretical sense, a grotesque amount of broccoli would have to be consumed for any effect on the thyroid to be noted. In any case, this was one bit of advice the president, having publicly expressed his distaste for the vegetable, was not likely to take. After those remarks, growers dumped truckloads of broccoli in front of the White House in protest, so he certainly would have had enough available had he chosen to indulge.

Graves' disease can be treated with drugs that block the synthesis of thyroid hormone, with radioactive iodide to deactivate the thyroid gland, or with surgery to remove part of the gland. This can result in hypothyroidism, or insufficient production of thyroid hormone, characterized by weight gain, lethargy, fatigue, constipation, cold, clammy skin, diminished sweating, thickened nails as well as coarse and prematurely grey hair. There can also be memory loss and intellectual impairment that may be wrongly ascribed to senile dementia. Hypothyroidism can also be due to a chronic inflammation of the thyroid gland that causes it to become enlarged and impairs hormone production. This is known as Hashimoto's disease. Whatever the cause of an underactive thyroid, the condition can be treated by oral thyroid hormones, with Synthroid being the most common commercial example.

For some unknown reason, thyroid disorders are five times more common in women than in men. In the West, newborns are tested for thyroid problems, but in India some 250 million people are estimated to suffer from iodine deficiency, which is manifested in decreased motor skills, low IQ and poor energy levels. Low levels of thyroid hormones are also believed to cause some ninety thousand stillbirths a year, and nine million children to be born with some form of mental retardation, often referred to as "cretinism."

In North America, salt is commonly iodized, meaning that it has potassium iodide added. But this does present a problem. Iodide is slowly converted to iodine in moist air. Since iodine is volatile, salt would slowly lose its power to protect against goiter. It is therefore common practice to add iodine stabilizers to iodized salt. These are substances that convert iodine back to iodide. Sodium thiosulphate used to be added, but it sounded too chemical and people became worried. So manufacturers switched to dextrose, which is as effective and sounds more innocuous. Sodium bicarbonate is also added because the oxidation of iodide occurs readily in an acid solution but not in a base. Disodium phosphate or sodium pyrophosphate are sometimes used to provide the alkaline conditions. These are also "sequestering agents," which bind trace metals that catalyze the oxidation of iodide to iodine.

Unfortunately, this easy protection against goiter is not being carried out everywhere. In India, because of the humid air, salt is usually sold in large crystals that are resistant to humidity. These are sometimes sprayed with potassium iodate (yet another substance capable of supplying iodine for thyroid hormone production), but that makes the salt look dirty, so people wash the crystals before crushing them. The consequences are tragic.

Frankincense is a tree resin that has long been used in incense and perfume production. Recent research shows that it may also be useful in the treatment of what medical condition?

Osteoarthritis. The wearing away of shock-absorbing cartilage that protects the ends of bones can result in stiffness and pain as bone rubs on bone. It is not primarily an inflammatory condition, meaning that it is not normally accompanied by swelling, heat or redness. But osteoarthritis can have an inflammatory component as bits of cartilage break off and cause a swelling around the joint. Since extracts of the *Boswellia serrata* plant, commonly known as frankincense, have anti-inflammatory properties, it comes as no surprise that they should have some efficacy in treating osteoarthritis. Frankincense has long been used in Ayurvedic medicine, mostly for digestive problems, and notably has not been associated with adverse reactions. Its use, however, has generally not involved standardized products.

Now, Laila Impex, an Indian pharmaceutical company, has developed a drug known as 5-Loxin from boswellia resin with a reproducible, standardized composition. The major active ingredient is acetyl-11-keto-beta boswellic acid, which is known to interfere with the activity of 5-lipoxygenase, an enzyme that catalyzes the formation of leukotrienes, which are compounds that promote inflammation. Of course, osteoarthritis sufferers want more than theory. And now they may have it: a randomized, double-blind, placebo-controlled trial using varying doses of 5-Loxin has shown evidence of benefit. Seventy patients completed the ninety-day study, with those taking 250 milligrams of the drug every day showing a significant improvement in pain score and stiffness. Furthermore, when

fluid was drawn from the subjects' knees, in those taking the medication, there was a significant decrease in matrix metalloprotein-
ase-3, an enzyme that can break down cartilage. While one cannot
make too much of a single study, there is room here for optimism,
given that no side effects were noted. And frankincense may have
yet another effect—at least in mice. Incensole acetate, a compound
found in frankincense smoke, relieves anxiety and depression. I'm
not sure how one diagnoses depression in a mouse, but that's another
story. In any case, those wise men bearing gifts of frankincense may
have been onto something.

A study showed that a daily intake of two
500-milligram tablets of quercetin, a flavonoid
found in apples, was effective in reducing the
symptoms of chronic prostatitis. If an average apple
contains 0.1 grams of quercetin, how many apples
would have to be eaten every day to match the
dose used in the trial?

Ten. Apples have a long-standing reputation of being a very
healthy fruit. And this reputation is actually based on more than
folklore. Epidemiological studies have linked apple consumption
with a reduced risk of lung cancer, heart disease, asthma and diabetes. No, you don't have to eat truckloads of apples to reap these
benefits—one or two a day can make a difference. Exactly which
of the more than three hundred naturally occurring compounds
in apples is responsible for the health benefits is unknown, but
flavonoids, based on their antioxidant and anti-inflammatory
properties, are strong candidates. Of these, quercetin has probably

been the most thoroughly investigated. For example, in a placebo-controlled trial of men suffering from chronic prostatitis, 67 per cent of the patients taking quercetin had a significant alleviation of symptoms, while only 20 per cent of the men in the placebo group showed improvement.

When male cyclists were asked to train at maximum intensity for three hours a day for three days, those who received a daily supplement of one gram of quercetin suffered far fewer chest infections than cyclists who were given a placebo. It is well known that athletes who engage in extreme physical activity are more prone to chest infections, as are soldiers who are under great physical stress during missions. This is probably because their immune systems are weakened and unable to fend off microbes effectively. Researchers believe that quercetin helps ward off infection by binding to viruses and bacteria and preventing them from replicating.

Quercetin may even be helpful for the mind. Although Alzheimer's is a complex disease, it is thought to involve free-radical damage to brain cells. Since quercetin has free radical–neutralizing capabilities, researchers at Cornell University decided to study its potential as a preventative for Alzheimer's by soaking rat brain cells in quercetin before exposing them to hydrogen peroxide, a free-radical generator. Quercetin proved to be very beneficial in protecting the cells from damage. Of course, rat brain cells in a petri dish are a long way from a functioning human brain, but still, the result is interesting. Nobody knows how many apples need to be eaten to offer such protection against Alzheimer's, but for sure, there is no downside to eating apples. And if you want the full benefit of quercetin, eat the peel, where most of it is found.

What naturally occurring chemical in the body would someone try to boost by taking 5-hydroxytryptophan?

Serotonin. Serotonin is one of hundreds of chemicals used by nerve cells to communicate with each other. These neurotransmitters are released by one cell, traverse the tiny gap called the synapse that separates nerve cells, and go on to activate an adjacent cell by binding to specific proteins called receptors, much like a key fitting into a lock. One of the most widely studied of these neurotransmitters is serotonin, a compound that plays a role in controlling our mental state. Antidepressants such as fluoxetine (marketed under the brand name Prozac) work by increasing the concentration of serotonin in the synapse. These drugs are called selective serotonin reuptake inhibitors (SSRIs) because they prevent the reuptake of serotonin by the cell that originally released it—which is sort of the body's way of recycling serotonin after it has done its job. Serotonin is produced in the body from the amino acid tryptophan, which is widely available in the diet. The metabolic pathway goes through an intermediate called 5-hydroxytryptophan, or 5-HTP, suggesting that, at least in theory, serotonin levels can be boosted either by ingesting tryptophan or 5-hydroxytryptophan.

Historically, there have been issues with tryptophan production, and a cloud was cast over this substance when a number of people died in 1989 from eosinophilia-myalgia syndrome, a condition that was traced to an impurity in a batch of tryptophan produced by one specific company. Because of this problem, focus shifted to boosting serotonin levels with 5-hydroxytryptophan, which can be extracted from seeds of the African *Griffonia simplicifolia* plant. A serotonin boost may do more than just elevate mood; it may serve as a sleep aid and migraine preventer. There is also some evidence that it can reduce cravings for carbohydrates and that it may alleviate the symptoms of fibromyalgia. The usual dosage of

5-HTP is in the range of one hundred to two hundred milligrams, one to three times a day. Evidence for efficacy is modest, but so are the risks. Cross-reactions with drugs are possible, particularly with those that also have an effect on serotonin levels, such as antidepressants, opiates and migraine medications. Needless to say, the use of 5-HTP should be discussed with a physician.

Why would a person's breath be tested for the presence of hydrogen gas?

To diagnose lactose intolerance. Lactose intolerance, the most common food intolerance in the world, is the inability to digest lactose, the sugar found in milk. Milk contains a fair bit of this sugar, roughly 5 per cent by weight. Lactose, the only carbohydrate of animal origin of any significance in the diet, is a disaccharide, meaning that it is composed of two smaller sugars, glucose and galactose. The breakdown of lactose in the digestive tract and the subsequent absorption of its components into the bloodstream require the presence of an enzyme called lactase. In the absence of this enzyme, lactose is not absorbed and its buildup draws fluid into the small intestine, causing diarrhea and often a very urgent need to visit the facilities. At the same time, the unabsorbed sugars serve as food for intestinal bacteria, which then produce gas as a byproduct. The buildup of gas can cause bloating, pain and flatulence. Indeed, the world record for gaseous emissions is held by a lactose-intolerant man who produced 141 outbursts in two hours after drinking two litres of milk. A great deal of this gas is hydrogen, some of which is absorbed into the bloodstream and is exhaled from the lungs. That is exactly why diagnosing lactose intolerance involves measuring the concentration of hydrogen in breath exhaled after a person drinks a standard dose of lactose.

Most cases of lactose intolerance are genetic. Populations that have not had milk as a mainstay of their diet have slowly lost the need to digest lactose. That's why about 90 per cent of Asians are affected, but only about 20 per cent of people of European origin suffer. But a lack of lactase can also be caused by intestinal diseases such as celiac or Crohn's disease. Some antibiotics—neomycin, for example—can also cause lactase deficiency. There are several approaches to dealing with lactose intolerance. Lactase can be added to milk to break down the lactose before it ever enters the body. Such milk will often be labelled as "lactose-free," "lactose-reduced" or "modified for easier digestion." This is a completely safe product.

Lactase drops, which can be added to milk, are also available, as are pills designed to be taken before indulging in milk products. In this case, the missing enzyme is replaced from an outside source. Avoidance of dairy products is an obvious measure for lactose-intolerant people, although most can consume small amounts of dairy products, especially yogurt, in which the fermenting bacteria use up most of the lactose as a source of food. Unfortunately, some people are so intolerant that they will even react to tiny amounts of lactose found as a filler in some medications. Such sensitive people can still drink fresh milk, as long as it comes from a bear. For some strange reason, bear's milk contains very little lactose. Of course, acquiring the bear's milk does present problems of its own.

What might a dermatologist call "liquid gold?"

Liquid nitrogen. It is so useful in dealing with various skin lesions, including some skin cancers, that its value can be likened to that of gold. Liquid nitrogen boils at −196°C, drawing heat from its surroundings in a dramatic fashion as it converts from a liquid to a gas. Tissues with which it comes into contact freeze instantly, as ice crystals form inside cells. Then, as the cells thaw out, the crystals expand and damage the cell walls, causing the fluid inside to leak out, which kills the cells. A very similar effect can be seen when a frozen steak thaws out. The liquid that was previously locked up in cells now escapes and reveals itself as a red juice.

Liquid nitrogen can be used to cool a metal probe that is then applied to the skin, but more commonly it is applied directly to the skin as a spray or by means of a cotton-tipped swab. Warts, keratoses and some skin cancers can be readily treated in this fashion. And they have been treated like this for more than a hundred years, ever since the production of liquid nitrogen became possible. Air, which of course is a mixture of nitrogen and oxygen, can be liquefied when put under great pressure. When liquid air is allowed to warm up, nitrogen boils off first and can be collected.

The key to using liquid nitrogen therapeutically, though, was the development of the Dewar flask, or Thermos bottle as some call it, which allowed super-cold liquids such as liquid nitrogen to be temporarily stored in a doctor's office. Dewar flasks can be filled from a liquid nitrogen tank, which is a thick-walled metal container capable of withstanding high pressures. That, of course, is necessary because nitrogen can only be stored as a liquid under high pressures. The storage tanks have to be equipped with a safety valve, so that if the pressure inside becomes dangerously high, as may happen if the temperature is increased, nitrogen can be vented as a gas.

Should the safety valve not function, the results can be dramatic, as evidenced by a spectacular accident at Texas A&M University where a not-so-bright graduate student in chemistry tried to fix a faulty release valve by clamping a brass fitting over it. The result of this folly was the nitrogen tank taking off like a rocket, completely penetrating the floor above and finally lodging in the attic. Luckily, the spectacular launch occurred in the middle of the night and nobody was hurt. Of course, it is not the liquid nitrogen that should be blamed, but the faulty handling. Liquid nitrogen can be used safely. You can even make ice cream with it: just blend together the usual components—the milk, eggs, sugar, cream and flavouring—and then, instead of using an ice cream freezer, just stir in some liquid nitrogen. Presto, a fresh batch of ice cream is ready to eat!

When is blood-coloured urine not worrisome?

If you've been eating beets. Beetroot gets its colour thanks to a family of compounds known collectively as betacyanins. Ingestion of beets can result in a bright red discoloration of the urine, and quite a scare for the urinator, who may confuse it with blood. Blood in the urine is a frightening prospect and a possible sign of serious disease, while the presence of betacyanins is benign and possibly even healthy. Interestingly enough, not everyone produces red urine after eating beets, a fact that leads many researchers to believe that production is genetically determined. Experiments have, however, cast doubt on this explanation. When subjects are given a fixed amount of beetroot to eat, and their urine is chemically analyzed, they all show the presence of betacyanins, but in some cases in amounts too small to impart a visual effect. When the dose is increased, subjects who were visually "non-excretors" begin to produce red urine. Furthermore, when subjects ingest the same amount of beetroot on separate occasions, they produce urine of varying shades of red. Factors other than genetics are obviously involved.

Times of planting and harvesting greatly influence betacyanin content, so beets purchased at different times may have different effects. Betacyanin colour also is dependent on acidity, being more stable as acidity is reduced. At the normal pH of the stomach, about 2, rapid decomposition of beet pigment occurs. If the acidity is reduced, such as by taking drugs for excess stomach acid, the chance of excreting red urine is increased. In one reported case, a gentleman who had never previously experienced the red urine phenomenon was scared out of his wits by the brilliant red colour he produced after a course of ranitidine, a drug used to reduce stomach acidity.

The presence of iron, in the form of ferric ions in the stomach and intestines, can also prevent the beetroot colourants from being absorbed into the bloodstream. Since iron can be complexed by oxalic acid or by ascorbic acid (vitamin C), these compounds, which are widely distributed in foods, can also determine whether eating beets will produce red urine. There seems to be enough evidence here to suggest that the production of red urine after eating beets is not controlled by genetics, but rather is a function of stomach acidity and the presence or absence of other dietary components.

According to researchers at the University of Wisconsin, not only is beet-induced red urine not a matter of concern, it may actually be healthy. It seems beet extracts can stimulate the liver's production of "phase II enzymes," which help eliminate toxins, including carcinogens. This property has been shown in mouse liver cell assays, but such experiments are known to be good models for what happens in the human liver. Eating beets may therefore provide some protection against cancer. Incidentally, turning beet-red in the face is solely a matter of embarrassment and has nothing to do with eating beets.

What dietary supplement derives its name from the Latin word for flesh?

Carnitine. This compound was first isolated from meat in 1905 and named after *carnus*, the Latin for flesh. Carnitine is needed to transport fats from the bloodstream into cells, where they can be burned for energy. Although carnitine is essential for health, it does not have to be supplied by the diet since it can readily be made by the liver and kidneys from lysine and methionine, two common amino acids. Because carnitine plays a definite role in the burning of fats, there has been interest in studying its potential to increase endurance, to reduce weight as well as to lower levels of cholesterol and triglycerides. Theoretically, carnitine could be of use in these areas, but the human evidence accumulated has been disappointing. More than twenty years of research using two to six grams of carnitine a day for extended periods has shown no benefit in athletic performance or in weight loss.

More positive results have been seen in patients with heart disease. The heart, of course, is a muscle, and it derives the energy it needs to function partly from burning fats. When it is deprived of oxygen—as during periods of angina, or after a heart attack—fat breakdown and energy production are reduced. In such cases, carnitine has been shown to be of some help. When given to heart attack victims for four to eight weeks at a dose of two to three grams a day, carnitine can reduce damage to the heart muscle. Angina sufferers may also be able to increase their exercise tolerance with a dose of about two grams a day. In some studies of people who have leg pain because of poor circulation (intermittent claudication), two grams a day has resulted in a significant improvement in the distance they can walk pain-free.

There is also some evidence that suggests that cancer patients who commonly experience fatigue from poor nutrition, radiation or chemotherapy may experience enhanced energy and improved mood with carnitine supplementation. A few studies have shown a relief of nerve pain associated with type 2 diabetes. The clearest indication for carnitine supplements is for kidney patients who are known to experience impaired production of the compound and who are also known to excrete significant amounts of carnitine. Because carnitine has only been clearly shown to be beneficial in conditions that require a physician's attention, it is only available in Canada by prescription. U.S. laws are more lax, and since carnitine can be found to occur naturally in meat, it is regulated as a dietary supplement. Hence it is commonly advertised as an energy-inducing substance and a weight-loss aid, in spite of a lack of clinical evidence.

Ultraviolet light carries sufficient energy to break chemical bonds, as is evidenced by the link between exposure to UV light and skin cancer. What technology uses the bond-breaking ability of ultraviolet light to protect the general population from disease?

Water treatment. Most people take water treatment for granted. Open a tap or turn on a shower and you expect clean water to emerge. Of course, in this context, the meaning of "clean" is somewhat elusive, since the increasing sophistication of chemical detection techniques means that what may have been considered clean just a few years ago can now be found to be tainted by chemicals. Whether the presence of contaminants at the parts-per-trillion level is significant is debatable, but there are some nasties, such as NDMA (N-nitrosodimethylamine), which may present a risk of cancer even at such trace levels. This substance is a byproduct of many industrial processes, but it also forms in water that has been disinfected with chloramines. Interestingly enough, municipal water treatment facilities have been moving to replace chlorine treatment with chloramine treatment because of the byproducts of chlorination. Compounds such as chloroform, formed during chlorination, are potentially carcinogenic. When chloramine is used instead of chlorine, these do not form. But we may be going from the frying pan into the fire, because chloramine treatment leads to NDMA formation, which may be a bigger problem than chloroform.

But there is a solution. Water can be passed through a reactor equipped with ultraviolet lights. The wavelength of UV used, in the 200-to-290-nanometre range, readily breaks bonds in NDMA and causes the molecule to self-destruct. And NDMA is not the only substance susceptible to the destructive effects of UV light. Waterborne bacteria such as E. coli and protozoa such as cryptosporidium and giardia can also be destroyed by ultraviolet light, and when the use of hydrogen peroxide is combined with UV treatment, pesticide residues, pharmaceutical products and endocrine disruptors found in personal-care products are also broken down. Furthermore, UV can also be used to treat waste water from sewage treatment to cut down on contaminants entering the environment. This then places less of a burden on municipal water treatment facilities. Another bonus is that ultraviolet treatment, unlike chlorination, leaves no worrisome residue in the water. It is not the end-all in water treatment, but it is an extremely useful technology to cut down on trace contaminants in our water supply.

Circe was the sorceress in Greek mythology who drugged Odysseus' crew to make them forget their homeland, then proceeded to turn the men into swine. When Odysseus set out to rescue his crew, he protected himself with an antidote derived from the snowdrop flower. This same antidote is used today in the treatment of what disease?

Alzheimer's. Myths often have some basis in fact. The ancient Greeks knew that an extract of the *Datura stramonium* plant, known today as thorn apple or jimsonweed, had the ability to rob people of their memories and cast them into a hallucinatory state sometimes accompanied by delusions of being turned into animals. The active ingredient in datura is atropine, which has the effect of blocking the action of an important neurotransmitter known as acetylcholine. Neurotransmitters are molecules that transmit information from one nerve cell to another and are responsible for effects ranging from controlling the heartbeat to the retention of memory. Alzheimer's disease is characterized by a deficiency of acetylcholine, and progress of the disease can potentially be slowed by drugs that increase the levels of acetycholine in the brain. One way to do this is by blocking the action of an enzyme known as acetylcholinesterase, which breaks down acetylcholine. As it turns out, the snowdrop, *Galanthus nivalis*, contains a natural acetylcholinesterase inhibitor known as galantamine.

This is what Hermes, the messenger of the gods, advised Odysseus to use to protect himself from Circe's drug-enhanced sorcery. And this is what peasants in the Balkan countries have used for centuries to treat themselves for various "nerve" problems. In the 1950s, a Bulgarian pharmacologist became interested in this folklore and alerted the medical community. Eventually, galantamine was isolated and tested in Alzheimer's patients. The results were not miraculous, but there was a definite slowing of decline and in some cases even improvement in cognition. Isolation from snowdrops proved to be difficult and the yield was low. Researchers, however, discovered that the common daffodil provided an excellent source of galantamine, and chemists also found a way to synthesize the compound in the lab. This, then, allowed galantamine to be extensively tested and paved the way for its appearance as Reminyl in the prescription marketplace. The name was eventually changed to Razadyne, following the deaths of two people who had been mistakenly given a diabetes medication, Amaryl (glimepiride), instead of Reminyl.

When Hermes introduced Odysseus to the snowdrop, he referred to the plant as the "moly," which just may be where our expression "holy moly" comes from.

In 1929, Philip Hensch at the Mayo Clinic observed that patients with rheumatoid arthritis improved greatly if they developed jaundice or if they became pregnant. This observation led to the discovery of what drug?

Cortisone. At the time, researchers had already determined that the adrenal glands, which sit on top of the kidneys, were involved in the body's response to stress. Back in the nineteenth century, Thomas Addison had noted a relationship between shrivelled adrenals and increased susceptibility to infection and had hypothesized that the adrenals must produce some substance that helps the body deal with stress. Hensch, aware of this theory, now suggested that jaundice or pregnancy stress the body, which in turn responds by increasing the activity of the adrenals. Tadeusz Reichstein in Switzerland and Edward Kendall at the Mayo Clinic eventually isolated a number of steroids from adrenal glands, one of which they identified in 1935 as cortisone.

It took another decade before chemists at the Merck pharmaceutical company were able to synthesize cortisone in sufficient amounts to experiment with therapeutic applications. They showed that in normal doses the compound was important in maintaining the activity of the immune system, but that in higher doses it *suppressed* immune activity, including inflammation. Finally, in 1948, Drs. Charles Slocomb and Howard Polley at the Mayo Clinic injected a rheumatoid arthritis patient, who had insisted on being a guinea pig for whatever experimental treatment was available, with one hundred milligrams of cortisone. The pain relief was almost miraculous. They then tried cortisone in a number of other patients, including one who had been totally bedridden. Upon treatment he got out of bed and attempted to dance! Another patient took seven baths in one day to compensate for the ones she had missed while suffering from arthritis. Unfortunately, there wasn't enough cortisone available to continue treatment, and the patients relapsed. But it was clear that a breakthrough had been achieved.

Today, cortisone and its chemical relatives, referred to as glucocorticoids, constitute an important class of drugs for the treatment of a variety of diseases. As the term implies, these drugs increase blood levels of glucose, the substance the body uses as a source of energy for tasks such as fighting off infections. Higher doses of glucocorticoids, however, have the opposite effect and *suppress* the immune system. Since inflammation is characteristic of increased immune activity, cortisone can serve as an anti-inflammatory agent. That's why it is useful in the treatment of inflammatory conditions such as arthritis, eczema, asthma and Crohn's disease. As always, there is a cost to be paid for the treatment. The increased level of glucose in the blood can lead to diabetes, and depression of the immune system can impair the response to attacks by viruses and bacteria. Weakening of bones, possibly leading to osteoporosis, is another possible side effect, as is the buildup of fatty tissues, causing a "moon face." But there is no doubt that glucocorticoids have had a major positive impact on the practice of medicine, and that Philip Hensch, Edward Kendall and Tadeusz Reichstein were well deserving of the Nobel Prize for Medicine that they shared in 1950.

Why has it become safer in recent years for people with compromised immune systems to walk through Trafalgar Square in London?

Because until recently, Trafalgar Square was carpeted with pigeons. The square is of course a great tourist attraction, and the feeding of pigeons used to be a popular activity for Londoners and tourists alike. But there was a problem: the birds' droppings were unsightly and corrosive to the neighbouring buildings, and the flock, which at its peak was estimated at thirty-five thousand, was regarded by many as a health hazard. And with good reason: a number of infectious diseases are associated with pigeons—or, more specifically, their droppings. The most common one, although still rare, is histoplasmosis, a fungal infection. It is also known as Darling's disease, although there is nothing endearing about it.

Histoplasma capsulatum is a fungus found in bird and bat droppings as well as in soil. When bird droppings are moved about, as when maintenance workers try to clean it up, the fungi can become airborne and can be inhaled. Most people who come in contact with the fungus don't ever realize they have been infected because their immune system deals with it effectively. But immune-compromised individuals—such as those infected with the HIV virus, or patients undergoing chemotherapy—are more susceptible to the ill effects of the disease. Symptoms usually occur three to seventeen days after being infected, with an average of twelve to fourteen days. While symptoms may vary from person to person, the fungus mainly attacks the lungs, causing respiratory symptoms that are flu-like.

Chronic histoplasmosis can resemble tuberculosis, so it is not easily picked up by a physician. As a matter of fact, 40 per cent to 70 per cent of chest x-rays of those who are infected show up normal. If the disease becomes disseminated, it can lead to multiple organ failure and death if left untreated. Even the eyes can be affected. Ocular histoplasmosis damages the retina and can cause loss of vision that is not unlike macular degeneration. It therefore comes as good news for the immune-compromised that the pigeon population of Trafalgar Square has been reduced. In 2000, the sale of bird seed was terminated, albeit not without controversy. Other measures were also introduced to discourage the pigeons, including trained falcons. Bird supporters and tourists kept feeding the birds, however, which prompted Ken Livingstone, the mayor of London at the time, to enact bylaws in 2003 to ban pigeon feeding within the square. On September 10, 2007, the bylaws were tightened, sealing an outright ban on feeding birds in the area of the square. Now there are relatively few birds in Trafalgar Square, which is used for festivals and hired out to film companies—activities that were not feasible in the 1990s. And, of course, the statue of Lord Nelson that dominates the square is grateful.

LET'S
MOTOR!

What would happen if you put a dozen sugar cubes into the gas tank of a car?

Nothing. Sugar does not dissolve in gasoline to any significant extent, meaning that the sugar cubes would simply sink to the bottom of the tank and stay there. It's true that when sugar is heated to a high temperature it turns into the sticky mess we call caramel, which indeed could clog up an engine—*if* the sugar were to get there. But it doesn't. It is also theoretically possible that, through physical agitation, the sugar cubes could crumble into grains of sugar that could be carried by the gasoline into the engine, where they would caramelize and cause havoc. Except that cars are built with filters in the fuel line to catch such grit!

Of course, the fact that sugar is insoluble in gasoline does not stop Internet hacks from fuelling rumours about wasting car engines by dumping sugar in the gas tank. One popular myth has a predatory criminal slipping sugar into women's cars and then following the victim until her car stalls in some remote area. You can guess the rest. If enough sugar were poured into a gasoline tank, the fuel filter would eventually get clogged and the engine could stall. Of course, you couldn't predict when this would happen. In a worst-case scenario, a tank might have to be removed to dump out the sugar.

In the future, though, we may be talking about putting sugar into the gas tank instead of keeping it out. Sort of. Several biotechnology companies are working on converting sugar derived from corn into gasoline through the use of genetically engineered bacteria. Glucose, a sugar readily available from corn, is of course already being fermented into alcohol to produce "gasohol." But producing hydrocarbon fuels from glucose could be a more attractive approach than making ethanol. Hydrocarbons release about 30 per cent more energy per litre when burned and could conceivably take less energy to produce. When ethanol is produced by yeast fermentation, it needs to be distilled to remove water. This process requires a great deal of energy that would not be expended if hydrocarbons could be produced from glucose. And they can be.

As part of their normal metabolic processes, bacteria churn out fatty acids—molecules that, like hydrocarbons, are composed of long chains of carbon and hydrogen atoms, although they also have some oxygen atoms at one end. The genes involved in the biosynthesis of fatty acids have been identified, and similar genes that code for the production of long hydrocarbon chains without the insertion of oxygen atoms have been either constructed in the lab or isolated from plant sources. These novel genes can then be inserted into bacteria such as E. coli, which will dutifully crank out hydrocarbons. These molecules would be just like the ones derived from petroleum, and furthermore would not be contaminated with sulphur compounds, as petroleum is. Far from ruining our cars as a consequence of being put into the fuel tank, sugar may actually fuel them.

What would you be doing if you were using a hydrometer to test for the amount of sulphuric acid present?

Measuring the charge in a lead-acid car battery. Oh where, oh where would we be without lead-acid batteries? On horseback or on bicycles. The battery that starts our cars, trucks and buses was invented by French physicist Gaston Plante back in 1859. The pertinent chemistry is relatively straightforward. Each cell in the battery contains a plate of lead and a plate of lead oxide immersed in a solution of sulphuric acid. In such a system, electrons will spontaneously flow from the lead to the lead oxide, thereby generating an electric current. In the process, both the lead and lead oxide plates convert to lead sulphate, the sulphate ions being supplied by the sulphuric acid.

As the battery is being discharged, the concentration of sulphuric acid consequently drops. Exactly the reverse reaction happens when a running car charges the battery. The lead sulphate that coats each electrode now reconverts to lead at the anode and lead oxide at the cathode, releasing the sulphate ions to once again form sulphuric acid. Therefore, the degree to which a battery is charged can be determined by the concentration of sulphuric acid in the cell. The easiest way to do this is with a simple device called a hydrometer, which basically measures the density of a liquid in comparison to the density of water.

A hydrometer is a weighted glass bulb that floats in a liquid, its depth of immersion determined by the density of the liquid. Archimedes' principle states that a solid suspended in a liquid will be buoyed up by a force equal to the weight of the liquid displaced. In other words, the lower the density of a liquid, the farther the hydrometer will sink. Since sulphuric acid is denser than water, the hydrometer will float higher in a fully charged battery. With time, a battery will lose its ability to hold a charge. This is because not all of the lead sulphate that builds up on the electrodes with each charging cycle converts back to lead and lead oxide; some slowly changes into a stable crystalline form that does not redissolve readily. Devices that send a strong electric pulse through the battery to dissolve the lead sulphate are available and can be used to bring a failing battery back to life.

There are also claims that a battery that no longer holds a charge can be revived by the addition of magnesium sulphate (Epsom salt). Supposedly, this destabilizes the crystal structure of the accumulated lead sulphate and allows it to dissolve. Some people swear by this process; others claim it doesn't work. Nobody seems to have carried out a controlled trial to shed light on the question.

Why may Bolivia be the country that benefits most from the proliferation of electric and hybrid cars?

Bolivia has huge deposits of lithium, a metal needed to make the lithium-ion batteries that are used in most hybrid and electric cars. Lithium-based batteries are already widely used in cell phones, iPods and laptops, but production of lithium carbonate, the form of the metal needed to make lithium-ion batteries, has to be increased to meet the emerging need to supply cars.

The brine found under Bolivia's Salar De Uyuni, the world's largest salt plain, holds roughly half the world's supply of lithium in the form of lithium chloride and lithium carbonate. This desert of salt is so large that it was one of the first sights seen from the moon by Neil Armstrong and Buzz Aldrin as they gazed at the earth on that historic July day in 1969. The Bolivian government recognizes the importance of its lithium deposits and has vowed that the country will not be exploited by Western powers. Lithium may be the oil of the future, and Bolivia's socialist president, Evo Morales, has clearly stated that his country will retain control of its natural resources and foreign companies will not be allowed to start mining for lithium.

Bolivia hopes to extract and process lithium carbonate and sell it on its own. A pilot plant has already been set up to do this. The operation is not a simple one, because the brine under the salt flats has to be pumped to the surface and its water content allowed to evaporate. Then comes the difficulty of separating the lithium from the potassium, magnesium and boron that are also present in the brine. This is a multi-step process that culminates in the addition of sodium carbonate to precipitate lithium carbonate, which is then purified and used in batteries.

Of course, we must remember that batteries only store energy, they do not create it. Charging a battery requires the input of energy, which is great if solar, wind or hydroelectric power is used, but not so good if the energy comes from coal-powered electric plants. Lithium-ion batteries at this point hold the greatest hope for a fast charge and long-term power production. The makers of the James Bond movie *Quantum of Solace* missed out on the possibility of a more intriguing plot than the one they came up with. Much of the film was set in Bolivia, with Bond battling a villain who wanted to control that nation's water supply. A fight over the lithium fields would have been more appropriate. Incidentally, the lithium carbonate used in batteries is the same substance prescribed for the treatment of bipolar disorder.

Why do manufacturers incorporate paraffin wax into the rubber used to make tires?

To protect the tires from the damaging effects of ozone. We normally associate ozone with protection, not with damage. And that is correct when the ozone is present in the stratosphere, where it can absorb ultraviolet light from the sun. Excessive exposure to ultraviolet light can, of course, cause skin cancer. But ozone found at ground level is a different story. It irritates the lungs and can cause inflammation of the bronchial tubes. This can be very serious for anyone suffering from respiratory problems, and indeed the American Lung Association links as many as fifteen thousand hospital admissions every year to ozone pollution. Ozone can also damage the rubber in tires, but it does not react with paraffin wax, which can therefore protect the rubber by coating it. There is always a thin layer of paraffin on the surface of a tire, and as tires age, more wax migrates to the surface to replace that which has been worn off.

Now then, where does ozone at ground level come from? Basically, us. We drive cars, and that's where the problem lies. A car's engine becomes very hot and produces enough energy to allow oxygen and nitrogen in the air to combine and form nitrogen dioxide. This is the brownish-coloured gas that is associated with automobile-induced smog. It is a respiratory irritant, but the real issue is that it breaks down to form nitric oxide and atoms of oxygen, which in turn react with oxygen in the air to form ozone. This gas reacts with fabrics to discolour them, causes damage to crops such as tomatoes and can also react with rubber—including the rubber used in condoms stored in many a wallet. Take care!

In 1851, at the Great Exhibition of London, an American inventor captured the public's attention with a display in which everything from furniture and curtains to books, umbrellas, jewellery and plates was made of the material he invented. Who was he, and what was the invention?

The inventor was Charles Goodyear, and everything in the display was made of vulcanized rubber, the type of rubber on which the world's automobiles now roll. The Great Exhibition was conceived by Queen Victoria's husband, Prince Albert, as a "true test of the point of development at which the whole of mankind has arrived." Albert was a great patron of science and technology and hoped that the fair would encourage friendly competition among nations along these lines, leading to further scientific advances. The magnificent Crystal Palace was built in London's Hyde Park to house the exhibition.

America was well represented by displays of ice-making machines, false teeth, a hand-cranked dishwasher and even a coffin equipped with a pump to remove the air so that its occupant could be preserved forever. Samuel Colt himself was on hand, urging visitors to test out his six-shooter, and Cyrus McCormick proudly showed off his reaping machine. But it was the somewhat eccentric Charles Goodyear who had put together one of the most captivating displays. Every item in the exhibit, from floor to ceiling, from Bibles to condoms, from eyeglass frames to hydrogen-filled balloons, was made of Goodyear's invention, vulcanized rubber. Visitors were greeted at the entrance by rubber plants in rubber pots.

Since childhood, Goodyear had been fascinated by rubber and was aware of the limitations of this strange exudate of the rubber tree. While it had a number of obvious uses based on its impermeability to water, it was saddled with the classic problems of becoming hard and brittle in the winter and soft and gooey in the summer. Goodyear was determined to solve these problems, and devoted his life and his money to its pursuit. A number of times, he ran out of funds and ended up in debtor's prison. Having no scientific background at all, Goodyear went about the business of inventing by trial and error. He tried mixing rubber with various substances, having some moderate success with magnesium carbonate, nitric acid and copper. Then one day he met Nathaniel Hayward, who, while working for a rubber company, had discovered that spreading sulphur on rubber eliminated its stickiness.

Goodyear bought Hayward's patent and began to add sulphur to his own concoctions. Then one day he accidentally dropped a mass of rubber treated with Hayward's sulphur solution on a hot stove. On trying to clean up the mess, Goodyear noted that the material was stretchy and refused to harden with cold or soften with heat. Goodyear did not realize it, but the heat had allowed sulphur atoms to link together the long chains of rubber molecules, yielding novel properties. Eventually, the term "vulcanized" was used to describe this newfangled rubber, after Vulcan, the Roman god of fire.

Although vulcanized rubber became, and remains, a widely used commodity, Goodyear never profited from his invention, mostly because he had to plow a great deal of money into fighting patent infringements. The Crystal Palace exhibit of 1851 was truly wondrous, as was the building itself, as described by a visiting English clergyman: "That majestic palace of iron and glass! A while ago, its pillars were coarse rude particles, clodded together in some deep recess of the earth, and its transparent plates were sandy masses, without beauty or coherence. How a little fire and a little art have changed them." He could have used similar words to describe Goodyear's discovery.

Eighty-four spectators were killed at the Le Mans automobile race in 1955, when a car flew into the

stands and caused an intense fire when its body ignited. What was the car's body made of?

An alloy containing mostly magnesium. The car had been built by Mercedes-Benz using an ultralight magnesium alloy that was far lighter than steel or even aluminum. Reduced body weight, of course, leads to better performance. On that fateful day, about two hours into the race, the car driven by Pierre Levegh collided with another and catapulted into the stands. The fuel tank ruptured and the escaping fuel caught fire, blanketing the body of the car with the flaming liquid. Magnesium ignites at a relatively low 600°C, a temperature easily reached by burning fuel. Once the metal catches fire, its temperature rapidly escalates to some 3,000°C, resulting in a veritable fireball—at close range, actually brighter than the sun.

Early photographers made use of this reaction. Back in 1864, Alfred Brothers of Manchester produced the first ever photographic portrait using artificial light, which was generated by a piece of burning magnesium ribbon. This technique was soon superseded by magnesium "flashlamps" that produced a very bright light when magnesium powder was blown into a flame with a rubber bulb. Since the powder had a greater surface area than the ribbon, it allowed for more contact with the oxygen in the air and produced an even brighter flame. Better yet was flash powder, a mixture of magnesium powder and the oxidizing agent potassium chlorate. A spark from a flint or a blasting cap ignited the magnesium powder, which then heated the potassium chlorate, causing it to release oxygen, which further fuelled the fire. A very bright light was produced, but the reaction released copious amounts of finely powdered magnesium oxide, often leaving the subjects gasping in the cloud of white dust. By 1893, the Frenchman Chauffour had eliminated the risk of exposure to magnesium oxide powder by placing magnesium ribbon inside a globe of pressurized oxygen, creating what was perhaps the first photo flashbulb.

Now back to Le Mans. The burning Mercedes-Benz was like a giant flashbulb, but one that kept burning. Firefighters first tried to put out the flame with water, which actually made the inferno even more wicked. When magnesium burns, it produces enough heat to decompose water into hydrogen and oxygen. The oxygen then fans the magnesium flame, and it can also recombine with the hydrogen released to cause an explosion. A further problem is that magnesium can react directly with water to form magnesium hydroxide and hydrogen gas, which can ignite. How, then, do you put out a magnesium fire? Obviously, not with water. Nor with carbon dioxide. Magnesium reacts with carbon dioxide to form magnesium oxide and carbon, releasing a great deal of heat. The way to put out a magnesium fire is with a Class D fire extinguisher, which sprays out finely powdered salt. The salt blankets the fire, cutting off contact with the oxygen in the air. Had this been available at Le Mans, lives would have been saved.

THE
EMBODIMENT
OF SCIENCE

What part of the body derives its name from the Latin for the breadth of twelve fingers?

The first part of the small intestine, the duodenum. The small intestine links the stomach to the large intestine, or colon. The duodenum emerges from the stomach and is the shortest part of the small intestine. But it is a very important part, because this is where most of the digestion of food takes place.

Digestion is the chemical process by which complex molecules such as proteins, fats and carbohydrates are broken down by digestive juices into smaller fragments that can be absorbed into the bloodstream through the intestinal wall. These digestive juices are composed of bile from the liver and gallbladder, as well as trypsin, lipase and amylase, enzymes that are produced by the pancreas. The pancreas also releases bicarbonate to neutralize the acid entering the duodenum from the stomach.

And how do the liver and pancreas know when food has entered the duodenum? Because the duodenum is also an endocrine organ, meaning it secretes hormones—chemical messengers that travel through the bloodstream to trigger action elsewhere in the body.

Cholecystokinin and secretin are the enzymes that the duodenum secretes to stimulate the flow of digestive juices. Secretin was actually the first hormone ever identified. Digestive enzymes are also available as oral supplements for medical purposes. Patients suffering from chronic pancreatitis, cystic fibrosis, cancer of the pancreas or irritable bowel syndrome can benefit from these supplements. The enzymes can be extracted either from animal glands or from plant sources. Enzymes that digest proteins, for example, are extracted from unripe papaya or pineapple, while pancreatin usually is derived from pork pancreas.

What material was used to make the first artificial hip joint?

Ivory. The first attempt to replace a worn hip joint was made by Dr. Themistocles Gluck in Germany in the 1880s. It was common at the time for joints to be destroyed by tuberculosis, and Gluck experimented with hip, wrist, knee and elbow joints made of ivory. It was a reasonable choice because ivory is tough, and its chemical composition, calcium phosphate, is similar to that of bone. At first, Gluck's patients did well, but since drugs to control infection were not available, the infection soon returned, destroying the bone to which the artificial joint had been anchored.

It wasn't until the 1960s that hip replacement became a routine procedure, if one can call any surgical intervention routine. The hip is a ball-and-socket joint, formed where the rounded upper end of the femur meets a cup-shaped part of the pelvic bone called the acetabulum. It is critical that the artificial joint be made of materials that do not release tiny particles as a result of wear and tear.

Such foreign particles can be attacked by the body's immune system, resulting in the release of enzymes that destroy adjacent bone cells, loosening the joint and causing ultimate failure.

The head of the femur is usually replaced by a titanium or cobalt-chromium alloy ball, attached to a stem made of the same metals, driven into the femur. Coating the titanium stem with a thin layer of calcium phosphate, which is a close relative of hydroxyapatite, the bone material, reduces the risk of provoking an immune response. Natural bone grows onto the calcium phosphate coating, binding the stem to the bone, making the use of cement unnecessary. So, interestingly, Dr. Gluck was on the right track with his ivory, which is also essentially calcium phosphate. The socket is replaced by one made of ultra high molecular weight polyethylene, which is very strong and resists the release of small particles caused by grinding, but even stronger synthetic materials are being explored.

What kind of physician would be most likely to handle ultra high molecular weight polyethylene?

An orthopedic surgeon. As described above, ultra high molecular weight polyethylene is the plastic used to manufacture implants such as artificial hip joints and knees. This material has very different properties from polyethylene, the plastic used to make those ubiquitous shopping bags as well as numerous other items ranging from bottles to hula hoops.

All polyethylene is made from ethylene, a gas that is derived from petroleum. The prefix *poly-* means "many" and refers to the fact that in polyethylene numerous ethylene molecules have been linked together into a long chain. Just how long this chain is determines the properties of the plastic. Regular polyethylene is composed of a few thousand units of ethylene, while ultra high molecular weight polyethylene has a couple of hundred thousand units per molecule. These long chains can be made to line up parallel to each other through a special process known as gel spinning, and the close interaction leads to a very strong material.

In order to achieve the high molecular weight needed, special catalysts known as metallocenes are used. The plastic produced is so strong that, when used to make artificial joints, it can stand up to many years of wear and tear inside the body. The material is virtually inert and does not provoke a rejection reaction by the body. It isn't perfect, however. After years of grinding, the joints do release microscopic particles that can provoke inflammation. Efforts are underway to reduce this possibility by making the original material more stable.

One of the problems is that the polyethylene has to be sterilized before being inserted into the body, and the usual technique involves electron-beam irradiation. While this destroys microbes, it also produces free radicals, which can then react with the plastic and weaken it. Because of its antioxidant properties, the addition of vitamin E to the plastic may be a solution. Ultra high molecular weight polyethylene can also be made into fibres that, on a weight per weight basis, are fifteen times stronger than steel, and are even stronger than fibres made from Kevlar, the material traditionally used to make body armour. Since ultra high molecular weight polyethylene is lighter and more flexible than Kevlar, it is now often used—under the brand names Dyneema or Spectra—in place of Kevlar in bullet-proof vests and helmets. It can also be used to manufacture strong ropes, bow strings, fishing lines, ski coatings, sails and cut-proof suits for short-track speed skaters. The production of polyethylene of various molecular weights by using specific catalysts and reaction conditions is an excellent example of how chemical ingenuity can help produce materials with the desired properties.

Heparin is an anticoagulant that is commonly used in the management of deep vein thrombosis, pulmonary embolism and various heart problems as well as in open heart surgery, organ transplantation and dialysis. Where does the name heparin come from?

Heparin derives its name from the Greek *hepar*, meaning "liver." The substance was first isolated from the liver of a dog by Jay McLean, then a medical student at Johns Hopkins University. McLean was very interested in research and wanted to get a year of laboratory work under his belt before formally starting his medical studies, convinced that once he became a surgeon, as was his plan, he would not be able to pursue his interest in physiology. McLean approached Professor William Howell, a noted physiologist at Johns Hopkins, and expressed his interest in working on a research project, without any remuneration, as long as he had a chance of getting some results within a year.

Howell had been trying to identify factors that caused blood to coagulate and had discovered that a crude extract of animal brain tissue did just that. He asked McLean to try to purify the extract and isolate the specific substance that caused blood to coagulate. Howell had believed it to be a fat-soluble compound called cephalin. So McLean's task was to isolate and purify cephalin, then test its ability to coagulate blood.

It turned out that heart and liver tissue were a better source of cephalin than brain tissue, and McLean found that an extract of these had a strong coagulant effect. But to his surprise, when the extracts were stored for a few months, not only did they lose their ability to coagulate blood, they had a marked *anti*coagulant effect! The deterioration of cephalin on exposure to air had been noted before, but the anticoagulant effect that now appeared was a novel discovery. Clearly, there was some anticoagulant present in the liver extract, the effect of which was masked by the strong coagulant effect of cephalin until the latter deteriorated.

Howell was very skeptical of McLean's finding until the clever student came up with a dramatic demonstration. He asked the laboratory assistant to bleed a cat, then he placed the blood in a beaker and added a batch of his newly discovered anticoagulant. He placed the beaker on Dr. Howell's desk and asked him to see how long the blood took to clot. It never did. Howell was sold, and he shifted the focus of his research to isolating the novel substance—heparin, as it was now called—in its pure form. While he had some success, he was never able to produce heparin in a reliable fashion. The small amounts that could be isolated from dog livers were expensive and sometimes toxic due to impurities. Finally, in the 1930s, pure heparin was produced from beef intestine by Dr. Charles Best, of insulin fame, at the University of Toronto's Connaught Laboratories. But the seed had been planted some twenty years earlier by a young medical student's taste for physiology.

In 1973, Dr. Eric Simon of New York University commented that his and others' research had been stimulated by reasoning that "God clearly didn't put those receptors there for people to shoot up heroin or to get pain relief from morphine." What discovery resulted from this research?

The presence in the body of endorphins—molecules produced by the pituitary and hypothalamus glands to produce pain relief and a sense of well-being. The name derives from "endogenous morphine," meaning internally produced morphine. A very appropriate description, because the identification of these compounds stemmed from the discovery that molecules isolated from opium, such as morphine, provided pain relief by stimulating receptors on nerve cells.

Receptors are specific protein molecules that are configured to bind to opiates. But why should the body evolve receptors for molecules found in a poppy that grows in the Orient? Perhaps, researchers theorized, morphine just happened to accidentally resemble some sort of molecule that the body itself produced to modulate pain. And in 1975, at the University of Aberdeen, Dr. Hans Kosterlitz and colleague John Hughes found such a molecule. In fact, they found two closely related molecules, both extracted from pigs' brains. They called them "enkephalins," from the Greek for "in the brain." The enkephalins turned out to be pentapeptides, molecules composed of five amino acids linked together. Later, longer chains of amino acids, all incorporating the enkephalin structure, were found to stimulate opiate receptors and were called endorphins. Today, the term "endorphin" is used to describe all the various peptides that have opiate-like activity.

Kosterlitz not only discovered endorphins, he also found that there are subtypes of opiate receptors, raising the possibility that there might be one kind of receptor for pain relief and another one responsible for addiction to opiates. This discovery triggered research into trying to find drugs that might fit one receptor without stimulating the other, perhaps leading to non-addictive opiates. Endorphins are also generated during orgasm and have been linked with the "runner's high." Some studies have even suggested that acupuncture needles stimulate the release of endorphins, based on the observation that naloxone, a drug that blocks opiate receptors, can negate the effects of acupuncture. There is also some evidence that the placebo effect is due to endorphin release. Patients who gain relief from pain after being treated with a sugar pill often see a return of the pain after being injected with naloxone.

What happens when 11-cis-retinal is transformed into 11-trans-retinal?

We see. Our complex sense of vision depends on a simple molecular transformation. Cis-retinal is a molecule stored in the retina, a layer of cells at the back of the eye responsible for converting a light signal into a nerve signal. When light hits this molecule, it changes shape— from one in which the chain of carbon atoms that make up the molecule have a kink in them to one in which they assume a linear arrangement. Retinal is embedded in a protein molecule called opsin, with the opsin-retinal combo being known as rhodopsin. When energized by light, cis-retinal is converted to the trans form, putting a stress on the opsin molecule that envelopes it. The result is that trans-retinal breaks loose from opsin, in the process altering the shape of the latter. It is these molecular contortions that stimulate the optic nerve to send a signal to the brain, and we see!

Trans-retinal, now disengaged from opsin, is then converted back into the cis form via a series of enzymatic reactions, after which it's ready to attach to another opsin molecule to again form rhodopsin. And the cycle continues. But where does 11-cis-retinal come from in the first place? Vitamin A. This substance, technically called all-trans-retinol, has to be acquired from the diet, either directly or in the form of beta carotene, which the body then converts to vitamin A. A deficiency in vitamin A can lead to visual problems, so the old adage about eating carrots to improve vision is not completely without merit. If the problem is caused by vitamin A deficiency, eating foods rich in carotenoids can indeed help.

In 1822, Alexis St. Martin was accidentally wounded in the abdomen by a musket. This allowed his physician, William Beaumont, to make a significant contribution to medical knowledge. What process did Beaumont investigate?

Digestion. William Beaumont had become an army doctor after an apprenticeship with Dr. B. Chandler in St. Albans, Vermont. When he was stationed at Fort Mackinac in Michigan, an accident occurred that allowed him to make the first systematic study of the process of digestion.

A young French-Canadian army porter, Alexis St. Martin, was wounded in the stomach when a musket accidentally discharged. He was brought to Beaumont, who was unable to close the wound. St. Martin developed an infection, for which, according to accepted practice at the time, he was bled by Beaumont. He survived the ghastly procedure only to be turned into a living laboratory. After about eighteen months the hole partially healed, becoming sort of a valve through which the stomach contents could be sampled. And sample Beaumont did, for about nine years!

He confirmed that the gastric juices were acidic, a property previously noted by the famous Flemish physician J.B. van Helmont in the 1600s. He also showed, as van Helmont had done before, that acidity was not enough for digestion because putting food into a straight acid solution did not lead to its breakdown. There must be some other substance secreted by the stomach that was critically important, he maintained. Soon after Beaumont's investigations laid the foundation, this critical substance was isolated and identified as the enzyme pepsin.

Beaumont also showed that juice removed from the stomach and placed in a glass jar was capable of digesting food, just the same as in the stomach. There was no "vital force" the human body possessed that was required for digestion, as some had maintained. Beaumont carried out hundreds of experiments using different foods, measuring the time taken for "chymification" in the stomach and in vials. He also measured the time it took for the stomach to empty after various meals. Alcohol, he discovered, irritated the lining of the stomach, and he therefore recommended that patients with heartburn stay away from such beverages. Whenever we take an antacid for an upset stomach, we should give a little thought to William Beaumont and his elegant work. He convinced doubters that stomach juice was not an inert liquid by resorting to facts, which he said "were more persuasive than arguments." In 1853, Beaumont fell off a horse and developed an infection. Unfortunately, unlike St. Martin, he did not survive.

CHEMICAL WARFARE

During the Second World War, the Germans attacked Britain with V-2 rockets that had a range of more than three hundred kilometres. What fuel did the V-2 use?

The V-2 was a sophisticated rocket that used alcohol as a fuel and liquid oxygen as the oxidizing agent. In the presence of oxygen, alcohol burns with a hot flame, producing carbon dioxide and water vapour. As these hot gases are expelled from the engine, the rocket is propelled in the opposite direction according to Newton's third law: For every action, there is an equal and opposite reaction. More than four thousand V-2 rockets were launched by the Germans against targets in western Europe and Britain, causing a great deal of damage when the explosives in the nose cone exploded on impact. The V-2s were designed by Wernher von Braun, the German rocket genius, and were built mostly by slave labour. The first U.S. rocket that boosted a man into space, the Redstone, was modelled on the V-2. In fact, the V-2 was the world's first spaceship, reaching an altitude of some one hundred kilometres as it sped toward its target.

Two ice cubes look identical and are both made from water. Yet when they are dropped into a glass of water, one floats and one sinks. What is the difference between them, and what implications does it have for warfare?

The ice cube that sinks is made by freezing "heavy water," which can be used in the production of plutonium for atomic bombs. The term "heavy water" can be taken quite literally—a molecule of heavy water is indeed heavier than ordinary water, but it is still water. How can this be? Everyone knows that a molecule of water is composed of an oxygen atom and two hydrogen atoms. But not everyone knows that hydrogen atoms are not all alike. They all have a single proton in their nucleus, as they must, since it is the number of protons in the nucleus that determines the identity of an atom. If there are two protons, it is no longer hydrogen and we have an atom of helium.

Atoms, however, can also contain another type of particle in their nucleus, known as a neutron. These have a mass identical to that of a proton, so they can change the mass of an atom without altering its identity. Atoms that differ only by virtue of having different numbers of neutrons in their nucleus are known as isotopes. An atom of hydrogen that has one neutron is called deuterium and will weigh twice as much as an atom of hydrogen that has no neutrons. Roughly one in every 6,400 hydrogen atoms in nature is a deuterium, meaning that some water molecules naturally contain a deuterium atom. What is their proportion to ordinary H_2O? Since each water molecule has two hydrogens, the chance of a water molecule having one deuterium in it is one in 3,200. The chance of having *two* deuterium atoms would be one in 3,200 squared, or roughly one in forty-one million.

Various sophisticated techniques exist that allow the deuterium oxide, or heavy water, molecules to be separated and produce pure D_2O. The question is why anyone would be interested in doing that. Sinking ice cubes make for a neat parlour trick, but hardly justify the expense of isolating D_2O from ordinary water.

Heavy water can do more than make ice cubes sink; it can slow down rapidly travelling neutrons. This is critical in some nuclear reactions, such as those used to generate energy—for example, in CANDU reactors. The name actually stands for Canada Deuterium Uranium, and is a type of reactor sold by Canada around the world. The use of heavy water allows these reactors to function without requiring enriched uranium. Most uranium atoms in nature occur as uranium-238, meaning they have 92 protons and 146 neutrons in their nucleus. But roughly one in every 140 uranium atoms has only 143 neutrons. This isotope is the most fissionable, can produce vast amounts of energy and does not require the use of heavy water in reactors. But separating it from the far more abundant U-238 is difficult. Thus the appeal of CANDU reactors, which can function with U-238. Unfortunately, heavy-water reactors can also be used to produce weapons-grade plutonium. That's why the production of heavy water is monitored by various agencies around the world. Still, it is not a problem for rogue governments to get their hands on sufficient heavy water to produce a bomb.

The Empire State Building is topped by a famous Art Deco spire. What was it designed to do?

Serve as a docking mast for dirigibles. Although the building was completed in 1931, during the height of the dirigible era, it was never used as a terminal for airships, as had been originally planned. Unlike balloons, dirigibles had a lightweight, cloth-covered aluminum skeleton that contained separate pockets of hydrogen gas. The first successful dirigible was engineered by the Prussian military genius Count Ferdinand von Zeppelin, who had purchased the plans for a rigid airship from the widow of Croatian inventor David Schwarz.

The first flight of a zeppelin, as the airships came to be called, took place in 1900, and within a decade, the *Deutschland*, powered by two diesel engines, was transporting paying passengers. Then came the First World War, during which zeppelins took on a different role. They flew reconnaissance missions spotting Allied ships in the Atlantic and were also pressed into service as bombers. The Germans built a total of eighty-four zeppelins, dropping close to six thousand bombs, mostly on England. Militarily, the zeppelins were not all that effective, but they did create a great deal of panic. Due to the impressive height at which they flew, there was little defence against the airships.

Many of the bombs they dropped were incendiary devices based on the chemistry of the thermite reaction. When a mixture of iron oxide and aluminum is ignited by means of a fuse, a tremendous amount of heat is produced, along with iron and aluminum oxide. The molten iron can set fire to any combustible substance it touches. After the war, dirigibles returned to peaceful use, although their military potential was kept in mind by both the U.S. and the Germans. The Americans, however, had something the Germans did not: a supply of helium, a gas that, unlike hydrogen, was not combustible and therefore far more suitable for airships.

The USS *Shenandoah* was the first ever airship that was held aloft by helium, the same gas used in today's blimps. Lacking helium, German zeppelins used hydrogen; the famous explosion of the *Hindenburg* in 1937 demonstrated the inherent problems with this gas. After the *Hindenburg* tragedy at Lakehurst, New Jersey, travel

by dirigible declined dramatically and was soon replaced by airplane travel. But the zeppelins, those behemoths of the sky—the
Hindenburg, for example, was almost the size of the *Titanic*—were
the world's first commercial aircraft.

What change was made to the production of the regulation U.S. Army bugle in 1942?

Plastic was substituted for the traditional brass. Metals, including
copper and zinc, the components of brass, were critical to the war
effort but were in short supply. The word went out to the U.S.
Army's Quartermaster Corps to try to find other materials that
could substitute for these metals in some applications.

Major E.L. Hobson had an idea: make bugles out of plastic!
Thousands of bugles were produced each year, consuming about
two pounds of copper per instrument. Could plastic be an adequate
replacement? Nylon had already made a name for itself as a replacement for silk in parachutes, and polyethylene insulation made radar
possible. When Elmer Mills of Mills Plastics in Chicago got a call
from Hobson asking about the possibility of producing a plastic
bugle, he was elated. After all, he had just produced a huge shipment of plastic toy bugles that sounded remarkably good. So he
happily took on the challenge of coming up with a version that
would meet the army's stringent requirements.

Mills figured that cellulose acetate, appropriately mixed with a
plasticizer, would do the job. This plastic had been known since
1865, when it was first made by reacting cellulose with a mixture
of acetic anhydride, acetic acid and sulphuric acid. With the help
of Frank Aman, an expert in designing woodwind instruments,

Mills came up with a cellulose acetate bugle that had all the right qualities. Mixing in a plasticizer, probably triphenyl phosphate, was the key. Plasticizers control the flexibility of a plastic, a property that is very important in allowing a bugle to produce the right musical tones. When a top-notch army bugler was asked to give the plastic bugle a try under field conditions, he was impressed. So were the officers who listened to his taps and reveille. And that very day, Mills got an order from the army for two hundred thousand plastic bugles!

Mills's bugles were the instruments that trumpeted victory in Europe and the Pacific. Although the cellulose acetate bugles performed very well, they did not last long. The plasticizer seeped out of the plastic, making for an unsightly white dust and leaving the plastic brittle. Today, the technology used to produce cellulose acetate is greatly improved, and the material is used to produce items ranging from tool handles and pen barrels to cosmetic containers and toothbrush handles. And, of course, toy bugles.

What berries are supposed to have helped British pilots shoot down German fighters during the Second World War?

As the story goes, British pilots used bilberries to shoot down German fighters. They didn't fire them out of their guns; they ate them—in the form of jam. This improved their night vision and made them more successful in dogfights.

Of course, this never really happened. It has been suggested that the British made up the story and spread it to distract the Germans from the real reason for the success of the British pilots, which was radar, but even this version seems to be an urban legend—accounts appear to be quite recent, and may have been cooked up by people attempting to market bilberries or bilberry extracts.

But there may be something to the benefits of bilberries. The rods of the eye, which we rely on in dim light, produce a chemical called rhodopsin. Without it, we cannot see in the dark. Well, it seems that bilberries contain compounds known as anthocyanins, which boost the production of rhodopsin. The anthocyanins may actually do more than improve night vision. They strengthen the tiny blood vessels that deliver oxygen to the eye. These capillaries, as they are known, sometimes rupture in diabetic patients and produce hemorrhage. In at least one study, diabetics given bilberry extract suffered fewer such hemorrhages.

There has even been some evidence that such extracts may slow the progress of macular degeneration, a terrible eye problem in which the central part of the retina slowly fails. Cataract risk may also be reduced. Since bilberry improves blood flow by strengthening capillaries, it may be of some use for people who bruise easily or who suffer from varicose veins or hemmorhoids. Bilberries are not usually available in Canada, but extracts are. These are often standardized in terms of their anthocyanin content. Normal doses range from two hundred to five hundred milligrams a day, taken twice a day (standardized to 25 per cent anthocyanins). People with diabetes and macular degeneration are probably the best candidates for taking bilberry extract. They may not become fighter pilots, but they may improve their odds in the fight against eye disease.

And what are bilberries? They're relatives of the blueberry and grow on small shrubs, mostly in Europe.

 ALTERNATIVE
(TO?) MEDICINE

The Lancet is a British medical journal with a huge worldwide readership and influence. What is the origin of the journal's name?

A lancet is a surgical knife classically used to open a vein for bloodletting, a procedure that dominated medicine for some two thousand years. Both ancient Chinese and Greek physicians bled their patients, but it was the reverence for Greek medicine that popularized bloodletting in Europe. The Greeks believed that illness was caused by an imbalance in the four humours—namely, yellow bile, black bile, phlegm and blood. Removing some blood, they maintained, would restore the balance and therefore health.

Lacking basic knowledge of how the body functioned, the humoral theory made sense to Europeans. Monks, and later barbers, freely bled the sick using either lancets or leeches. The red stripe on the barber's pole, symbolizing blood, can be traced back to that time. In the United States, Dr. Benjamin Rush, a leading medical light and signer of the Declaration of Independence, was a staunch supporter of bloodletting, sometimes bleeding as many as a hundred patients a day. He was convinced of the efficacy of the procedure probably by a combination of misinterpreting the list-lessness in patients who had lost blood as a sign of improvement, and selective memory. Rush remembered the patients who recovered, probably in spite of the treatment, and conveniently forgot those who had succumbed to bloodletting.

One of these unfortunate victims was George Washington, whose death, probably from some sort of infection, was precipitated by vigorous bloodletting. The demise of the former president in 1799 resulted in some doctors arguing that the barbaric procedure harmed more than it helped. Indeed, two years earlier, William Cobbett, a muckraking journalist, had accused Benjamin Rush of "contributing to the depopulation of the earth" after assembling data that indicated an increase in the death rate among patients of doctors who, at Rush's urging, had taken to bleeding victims of yellow fever. Rush retaliated by suing Cobbett for libel and won a huge settlement of $5,000—ironically, on the same day Washington died. He had triumphed essentially because of his reputation as an icon of medicine. The facts didn't matter.

Bloodletting continued unabated, with France importing more than forty-two million leeches as late as 1833. It wasn't until 1836, when Dr. Pierre Louis published his work *Researches on the effects of bloodletting in some inflammatory diseases*, that bloodletting began to wane. Louis had kept meticulous statistics on pneumonia patients at Paris's Hôpital de la Charité and concluded that those who were bled fared worse. Interestingly, back in 1809 a Scottish military surgeon, Alexander Hamilton, had carried out a proper clinical trial on sick soldiers, randomly dividing them into groups that would either be bled or not. The results clearly showed an increased death rate with bleeding—but Hamilton never published his results! We now know of the experiment because of the discovery of his personal documents in 1987. In fact, Hamilton could have published his results in *The Lancet*, which was founded in 1823. Some misery could certainly have been saved. But today, bloodletting still serves as a shining example of how longevity of a practice cannot be equated with efficacy.

What herb did the legendary Pied Piper of Hamlin supposedly use to get rid of rats?

Valerian root. Greek physicians in the first century called it "phew," so you can imagine what it smells like. Cat urine comes to mind. The dried roots of the valerian plant produce a really disturbing smell, at least to humans. But rats apparently are enchanted by it.

According to legend, the Pied Piper of Hamlin was hired in the thirteenth century by the elders of the town to get rid of rats. He used valerian to do so. The townsfolk should have paid him properly for his services but did not. So he charmed their children away, never to return. He sure didn't do *that* with valerian.

The plant, also known as the garden heliotrope, has long been used to induce sleep and relieve anxiety. Its tranquilizing effects have been compared with those of the benzodiazepines, such as Valium. There is, however, no chemical similarity between the active ingredient in Valium and the compounds found in valerian root. But these compounds—valeranone, valerenal, valeric acid and others known as valepotriates—seem to work in the same way as Valium. The mixture of chemicals in valerian root increases the levels of gamma-aminobutanoic acid (GABA) in the brain by blocking an enzyme that normally breaks this important neurotransmitter down. It seems, therefore, that the anecdotal evidence about the tranquilizing properties of valerian root may have some scientific basis. Maybe that physician in the eighteenth century who wrote that valerian had a "remarkable effect in quieting the nervous agitation which prevents sleep in delicate and irritable females" was onto something. But scientific studies suggest that he was not onto a whole lot.

While there is a lot of anecdotal evidence for improved sleep, objective measurements do not bear out the claims. In one French study, 89 per cent of people reported improved sleep, but examination of brain waves, which are good indicators of sleep quality, does not jibe with this claim. A number of clinical trials involving people who have sleep problems have shown that, on average, valerian didn't bring sleep on sooner, didn't reduce awakenings and did not diminish time spent awake. That does not mean that some individuals did not benefit.

Europeans certainly believe in the benefits. Valerian as powdered root is often prescribed by physicians for anxiety. It is also available as a tincture or as a powdered extract. You can also brew a tea from the root, but that is an unnerving experience because of the smell. In rare cases, people have become addicted to valerian, and withdrawal has produced rapid heart beat and tremors. Valerian should not be taken together with other tranquilizers such as Valium. But for mild anxiety, it may be worth a try. At least one company that manufactures steam irons thinks so. Their scientists are working on a version of valerian that can be added to steam irons to make ironing a brand new experience. Just inhale the vapours, relax and slow down. There should be no side effects, except perhaps attracting rats with the smell. And if rats appear, well, take some valerian. Or a Valium.

According to legend, an ancient Chinese goat herder noticed increased sexual activity in his flock

when the animals were grazing on a certain plant.
What was that plant?

The appropriately named horny goat weed. Whether the story
about the goat herder is true or not, we'll never know, but what we do
know is that traditional Chinese medicine has long used the plant
to increase libido in men and women and to improve erectile function in men.

Horny goat weed belongs to a family of perennial plants known
as Epimedium, which grow in the wild and are rarely cultivated. The
leaves are picked and used to prepare various extracts sold with a
promise of increasing a sagging sex drive. For men, an increase in sex
drive is not much good unless crucial sagging anatomical parts can
also be resurrected. And horny goat weed extracts promise to do
that as well.

Traditional use is not proof of efficacy, but thousands of years
of apparent satisfaction by patients of practitioners of traditional
Chinese medicine suggests that the chemistry of the plant is worth
looking into. And it has been investigated. Horny goat weed, like any
natural plant product, contains numerous compounds, but one in
particular stands out in terms of potential effects on sexual function.
Icariin has been extracted from the leaves of the plant and studied
in the laboratory. The effects have been, let us say, rather uplifting.

When injected directly into the penis of a rat, icariin causes an
erection. But what does it do for humans? So far there have been
no controlled double-blind trials, but there is some interesting evidence from laboratory studies to suggest that icariin has physiological activity similar to that of sildenafil, better known as Viagra.
As shown by researchers at the University of Milan, icariin, at least
in the test tube, inhibits the activity of human phosphodiesterase-5,
the enzyme that degrades cyclic guanosine monophosphate, the
compound responsible for increasing blood flow to the penis and
causing an erection.

The Italian researchers have also shown that chemical modification of icariin can lead to more active compounds. Of course, all of this doesn't mean that horny goat weed products sold in health food stores are effective; as with most such supplements, their exact composition is a matter of mystery. But eventually, standardized extracts of horny goat weed—or, more likely, some derivative of icariin—may provide an effective alternative to Viagra. Needless to say, as with any drug, natural or synthetic, possibilities of side effects exist. If icariin behaves like sildenafil, it is likely to have a similar side-effect profile as well, meaning that the drug may not be appropriate for some people taking heart medications such as nitroglycerin.

Why did King George VI, Queen Elizabeth's father, name one of his racehorses Hypericum?

The king had been so impressed with a homeopathic remedy derived from the plant *Hypericum perforatum* that he named the horse after it. The royal family's infatuation with homeopathy continues to this day in spite of a lack of scientific evidence attesting to the effectiveness of this curious practice.

Why is it curious? Because homeopathic remedies contain essentially nothing. Homeopathy was the idea of a well-meaning German physician, Samuel Hahnemann, who in 1790 offered up the thesis that a substance that causes certain symptoms when given to a healthy person in a high dose will, when given in a smaller dose, cure sick people who suffer from the same symptoms. He then compounded his illogical idea by suggesting that the smaller the dose, the more powerful the remedy. Indeed, the most powerful doses were the ones that had been diluted to the extent that they didn't even contain a single molecule of the original material. And that original material could be anything ranging from arsenic to bedbugs. Of course, because the diluted product contained no remnant of the substance, there certainly was no issue with toxicity.

How nonexistent molecules can have a therapeutic effect is a mystery. But the mystery only exists if the existence of a therapeutic effect is accepted. And the large number of placebo-controlled trials that have been carried out have shown no effect greater than what one would expect from a placebo. So, if homeopathic medications do not do anything, how did they gain such popularity in the nineteenth century? Essentially, *because* they do nothing. Regular physicians at the time did not have a stellar reputation. As Voltaire wrote, "Doctors are men who prescribe medicines of which they know little, to cure diseases of which they know less, in human beings of whom they know nothing." Unlike the bloodletting, blistering and purging offered by most physicians, homeopaths at least did no harm. It was a lot more pleasant to be given a dilute solution that contained nothing than to be subjected to bloodletting.

As medicine emerged out of the darkness in the twentieth century, homeopathy faded but did not die out. Scientific medicine has many limitations, and there is always plenty of room for placebo treatments. Which is likely what King George benefited from. His heirs, like Prince Charles, have sustained the belief in homeopathy, but I suspect that, should His Royal Highness come down with pneumonia, he would choose penicillin over hypericum.

You need flower petals, spring water, sunshine and a glass bowl. What are you trying to make?

A Bach flower remedy. Bach flower remedies have been around for close to a hundred years and were the brainchild of Edward Bach, a British physician. Actually, there doesn't seem to have been much brain involved in the development of this curious alternative healing method.

Bach was a traditionally trained physician who became disenchanted with the way medicine was being practised and began a search for novel healing methods. Then, in 1930, at the age of thirty, as he was walking through a field of flowers glistening with dew, he had an epiphany. He somehow surmised that the spiritual essence of a flower was transferred to the dew when the flower was exposed to the sun, and that this dew had healing properties.

This remarkable insight came to Bach, as he maintained, through "inspiration." He found that, to sense a flower's specific therapeutic potential, all he needed to do was hold a petal in his hand. Bach then went on to develop his healing essences by exposing flower petals floating in a glass bowl filled with spring water to sunshine. He claimed that in this fashion the flower's spiritual energy was transferred to the water, a few drops of which could then be used for healing purposes. Bach's bizarre notion was that the spirit of the plant communed with the human spirit and alleviated negative moods and the "lack of harmony" between the soul and the body that causes disease. Illness, Bach maintained, "will never be cured by present materialistic methods, for the simple reason that disease in its origin is not material, but is the result of conflict between the Soul and the Mind and will never be eradicated except by spiritual and mental effort."

Different flower essences are used for different purposes. For example, wild oat essence directs the confused or lost individual toward his or her life path. This, it is said, is the perfect remedy for someone with the "seeker" personality type to ease his soulful yearnings and tiresome wanderings. Wild oat is also recommended for youths seeking a vocation or anyone experiencing a mid-life crisis.

So, where is the proof for such claims? The marketers of Bach remedies say that they have no interest in proving that the remedies work; they let consumers make up their own minds. But others have carried out placebo-controlled trials, and these showed that all subjects, whether in the Bach flower essence group or the placebo group, experienced a decrease in anxiety, with no difference between the groups. The conclusion is that Bach flower remedies are an effective placebo for anxiety but do not have a specific effect.

So, if you are using "Five Flower Rescue Remedy" to ease fear and restore a state of calm and confidence, you are actually getting an inconsequential dose of flower extract with a heaping dose of placebo. Of course, I may only have this opinion because I'm not taking any beech flower essence, which "helps lessen one's tendency to be judgmental toward others or hypersensitive to their environments. Critical and blaming natures are often an indication of inner feelings of vulnerability and insecurity. This essence neutralizes intolerant and critical attitudes with feelings of tolerance and acceptance." I wonder which essence offers a cure for nonsensical thinking?

Extracts of bitter orange are being marketed as a legal replacement for what banned substance?

Ephedra. Excess body weight is a huge problem in North America and offers ample opportunity for marketers who claim to have an easy solution to the problem. Extracts of the ephedra plant were enthusiastically and successfully marketed for many years as a "natural" way to lose weight, until the U.S. Food and Drug Administration (FDA) put its foot down in 2004 and banned ephedra products, claiming that they posed an unacceptable risk of cardiovascular disease.

The supplement industry, which had been profiting heavily from ephedra, now put is weight and considerable muscle behind a legal challenge to the ban, and managed to get it reversed. There was, however, a proviso: the maximum daily dose in any marketed product was to be no greater than eight milligrams of ephedrine, the active ingredient in ephedra extracts. Canada allows up to thirty-two milligrams a day as long as no single dose contains more than eight milligrams. Even though ephedra in limited doses can now be marketed both in the U.S. and Canada, the controversy generated by the FDA's action took a toll on sales, especially when it became clear that the doses allowed were less than what had been shown in some clinical studies to have an effect on weight control.

The idea behind ephedrine as a weight-loss product is not bogus. Ephedrine is chemically similar to adrenalin, produced by the adrenal glands, and like adrenalin it can rev up metabolism. This essentially means that body fat is more readily broken down into substances that are readily eliminated from the body, generating heat in the process (thermogenesis). But with the image of ephedrine now tainted by health concerns, the supplement industry looked for a novel "natural" replacement and found it in the form of bitter orange extract.

Bitter orange, or *Citrus aurantium*, is, as the name implies, a member of the citrus family and is sometimes used to make marmalade but is rarely eaten as a fruit. It contains parasynephrine, which indeed bears a close chemical resemblance to ephedrine, justifying, at least in the eyes of its beholders, its marketing as an alternative to ephedra. The suggested dose is usually an extract containing anywhere from four to twenty milligrams of parasynephrine a day. This seems to be pure guesswork because no clinical studies have been performed to show that bitter-orange extracts can affect weight loss. But based on its chemistry, parasynephrine can be expected to have cardiovascular effects similar to ephedrine—namely, hypertension and irregular heartbeats. Furthermore, bitter-orange extracts can represent an even more significant cardiovascular risk for anyone on a monoamine oxidase type of antidepressant medication. Such drugs block the breakdown of parasynephrine and exacerbates its potential harmful effects. At this point, the weight of evidence is against bitter orange being a useful substance for weight control.

Black cohosh has been promoted as a remedy for what condition?

Menopausal symptoms. Black cohosh is a plant native to North America that has traditionally been used to treat menopausal symptoms. The evidence of its efficacy is mostly anecdotal. Dr. Katherine Newton of the Group Health Cooperative in Seattle decided to put black cohosh to a scientific test and enlisted 351 menopausal women between the ages of forty-five and fifty-five who were experiencing at least two hot flashes a night. The women were randomly assigned to one of five treatment groups: 160 milligrams of black cohosh daily; a multibotanical supplement containing 200 milligrams black cohosh and nine other herbal ingredients including alfalfa, pomegranate and Siberian ginseng; a multibotanical supplement plus increased soy consumption; hormone therapy; or placebo capsules. After three, six and twelve months, black cohosh was no better than placebo in reducing the frequency or severity of hot flashes or night sweats. The same was true for the other herbal products. Women who were given hormone therapy, on the other hand, had significantly fewer hot flashes and night sweats than women given placebo.

Another issue with black cohosh is its potential to interfere with the effectiveness of drugs used in cancer therapy. And then there is the problem that women relying on black cohosh may not be getting what they think they are getting. When researchers analyzed samples of eleven of the most popular black cohosh tablets and capsules available in New York City using a process called high-performance liquid chromatography, they identified hundreds of different compounds within each. Three of the products didn't have black cohosh at all, but instead contained an Asian species of actaea, a Chinese herb related to black cohosh but without any proven effects in easing menopausal symptoms. Supplement manufacturers substitute actaea for black cohosh because it is less expensive to produce. What about women who are happy with black cohosh? Any concern here? There is in the U.K., where health authorities want a warning label on black cohosh products because of the possibility of liver damage. This problem isn't particularly well documented, but certainly any physician finding liver dysfunction should ask the patient about the use of black cohosh—or, indeed, of any herbal remedy.

SCIENCE
MAKETH THE
CLOTHES

Modal fibre is advertised as a super-soft natural fibre ideal for underwear, bathrobes, towels and sheets. What "natural" substance is it made from?

Cellulose derived from beech trees. Modal is a type of rayon made from regenerated cellulose. While it is an excellent fibre and really does produce textiles that, unlike cotton, remain soft even after washing in hard water, do not pill and are resistant to shrinkage and fading, modal is not exactly natural. It does originate from naturally occurring cellulose, the main structural component of plants, but extensive chemical processing is required to produce the final result.

Back in 1894, Charles Fredrick Cross, Edward John Bevan and Clayton Beadle came up with what they hoped would be a cheaper substitute for silk. They were familiar with earlier attempts to mimic silk, particularly by Count Hilaire de Chardonnet, who, in 1884, had managed to treat cotton with nitric and sulphuric acid to produce "Chardonnet silk," which was indeed silky and attractive but was highly flammable. Workers, perhaps understandably, referred to it as "mother-in-law" silk.

Cross, Bevan and Beadle tried a different approach. They found
that cooking wood pulp in a solution of sodium hydroxide and
carbon disulphide yielded a highly viscous solution they called vis-
cose. Cellulose from the pulp had reacted with these chemicals to
form a new substance called cellulose xanthate. This viscous solution could then be passed through a spinnerette, a device with
many small holes resembling a shower head, to form long threads.
Immersing these in an acid solution regenerated the original cellulose, but now in the form of a fibre.

Eventually, these fibres were named rayon because they glis-
tened in the rays of the sun. Today, the technology for making
rayons like Modal is more sophisticated than in the early days,
but the concept is the same. Much is made of the fact that birch
trees, unlike the petroleum used to make synthetic fabrics, are a
renewable resource, and that their cultivation does not require
pesticides. Compared with cotton, growing the trees requires
less water and produces about ten times as much cellulose per
hectare. Still, the production of Modal requires extensive use of
chemicals, although modern processing methods ensure that very
little of these escape into the environment. So, while the use of
the term "natural" to describe Modal fibre is somewhat debatable, there is no doubt that clothing items made from this fibre
are very soft and comfortable. No wonder Victoria's Secret, La
Senza and Calvin Klein all promote the next-to-the-skin com-
fort of Modal.

**Who would be most likely to use the expression
"anything but cotton?"**

Long-distance runners or other athletes involved in any activity that produces sweat. Perspiration is the body's way of cooling itself. It takes heat to convert a liquid into a gas, and in the case of perspiration the heat needed to do this is supplied by the body. Evaporation of the moisture in sweat therefore causes a drop in the body's temperature. But should a layer of cotton be next to the skin, the sweat doesn't evaporate; instead, it's absorbed by the fabric. Not only does this prevent cooling, but the clothing becomes heavier because of the absorbed water.

Cotton is particularly adept at absorbing moisture because its cellulose molecules contain lots of oxygen atoms that have a strong affinity for the hydrogen atoms in water. This hydrogen bonding, as it is called, is responsible for cotton's absorbent properties. Absorption of moisture by clothing is exactly what one does not want if cooling is to be maximized. Rather, we look for a fabric that can wick moisture away from the skin and transport it through the material to the outer surface, from where it can then evaporate.

Actually, it is more the spaces between the fibres of the fabric rather than the fibres themselves that accomplish the wicking effect. What is involved here is a phenomenon known as capillary action. It can be observed most readily when a thin glass tube is dipped into water and the liquid is seen to rise up the tube, apparently defying gravity. In fact, the mysterious action is due to the difference in the attraction that water molecules have for each other, known as the surface tension, and the attraction they have for the glass surface. The molecules are more strongly attracted to glass, so those near the glass tube are drawn up it. A similar effect occurs when paper towel is dipped into water. There are numerous channels between the paper fibres that act like tiny tubes. Again, since the water molecules are attracted to the paper more than to each other, the liquid rises up the towel. If soap is added to the water, the effect is even more noticeable, since the soap molecules get in between the water molecules, decreasing their attraction for each other. In other words, soapy water can wet a porous surface like our skin or clothing even better, which is one of the reasons that a soap solution is more effective at cleaning than plain water.

A wicking fabric essentially works like a paper towel. But in this case, we want minimal absorption of moisture as it travels through the channels between the fibres, so we look for a material that is hydrophobic, or water repellent. Polyester, nylon and polypropylene fit the bill. These polymers can be drawn into very thin fibres that can be woven into fabrics with lots of channels between the fibres for maximum capillary action. Sweat is drawn away from the skin and is quickly transported to the outside of the fabric from where it evaporates, leaving the wearer cool and comfortable.

What invention was supposedly inspired by the inventor watching a cat clawing at a chicken in a cage and ending up with a paw full of feathers?

The cotton gin. The original cotton gin, as invented by Eli Whitney in 1793, was a wooden box fitted with a screen and a series of hooks on a drum. When the drum was rotated with a crank, the hooks pulled the fibre off the seed pods and through the screen, just like the cat defeathering the chicken. It didn't take long for the device to be scaled up and mechanized. Instead of a hand crank, horses or a water wheel were used to turn the drum. The impact of the cotton gin on the industry was tremendous, enabling about fifty times as much cotton to be produced in a day. This also meant that the demand for growing cotton increased, with an awful consequence: more slaves were needed to grow the crop. Cotton picking was tough work and had to be done by hand.

When cotton-picking machines were introduced, the labour problem was alleviated, but other issues had to be dealt with. The leaves
of the plant made picking by machine difficult, so chemicals that
removed the leaves were introduced. Magnesium chlorate was one
of the first to be used to make the plants shed their leaves, but
today's cotton growers have a wide array of defoliants available.

Why would a baby be wrapped in a BiliBlanket?

To treat jaundice, a condition caused by the liver's inability to remove
bilirubin, a breakdown product of red blood cells. A BiliBlanket
incorporates a flexible fluorescent panel that emits blue light, which
converts bilirubin into a form that is more readily excreted. The
livers of newborn babies are often not fully functional, resulting in
the accumulation of bilirubin in the body, which in turn leads to an
observable yellow appearance. Excessive accumulation of bilirubin
can cause extensive damage to the nervous system.

   In the 1950s, a British nurse who believed in the beneficial
effects of fresh air used to take jaundiced babies outside when the
weather permitted. A pediatrician, upon examining one of these
babies after an outing, noted a strange yellow patch on the abdomen that contrasted dramatically with the rest of the skin, which
now was a normal colour. It was quickly determined that the patch
was caused by a corner of a blanket which had fallen across the
baby's stomach. The conclusion was that somehow the sunshine
had cured the jaundice on the exposed skin.

   Research showed that it was the blue component of visible light
that was responsible for the curing effect, and today newborns
suffering from jaundice are routinely bathed in high-intensity

blue light. A normal liver can solubilize bilirubin and excrete it in the bile. In a liver that is not functioning fully, this solubilization does not occur, and the bilirubin begins to accumulate. Blue-wavelength light has the effect of changing the shape of the molecule, rendering it soluble and excretable.

In the past, infants were exposed to overhead fluorescent lights, requiring them to be isolated for long periods of time, with protection pads to shield their eyes from the light. The development of flexible fluorescent panels allows babies to be wrapped in a BiliBlanket, eliminating the need for eye protection. The blanket also allows the parent to hold the child during part or all of the therapy.

At first, DuPont considered naming the new substance Duparon, but eventually they settled on a name that is now recognized around the world. What is that name?

Nylon. Duparon, the name that was almost chosen, was a clever acronym for DuPont Pulls A Rabbit Out of Naphtha. Indeed it did. In the late 1920s, the chemical company made the decision, very innovative for the times, to invest in fundamental research. Chemists would be hired to pursue their theoretical interests instead of being asked to work on improving existing materials. The hope was that giving free rein to clever scientists would pay off in practical benefits. And did it ever!

Wallace Carothers, who had been lured away from Harvard University, was captivated by the idea of constructing giant molecules by joining together small ones, in much the way that links are made into a chain. By 1930, the idea had borne fruit, as Carothers and his associate Julian Hill produced the world's first polyester fibre. Unfortunately, it had a low melting point, which prevented it from being laundered or ironed. But with the principle of linking small molecules having been established, the search was now on for an improved fibre.

By 1935, Carothers's group had found a way to react diamines with diacids to make the material that would eventually be christened nylon. Since the raw materials needed were isolated from naphtha, a liquid distilled from coal tar, DuPont really did pull a rabbit out of naphtha. A rumour at the time had it that one of the components used to make nylon was pentanediamine, also known as cadaverine, which was isolated from decaying dead bodies. This, of course, is not true, but pentanediamine, which really is found in decaying flesh, can also be derived from naphtha, and was mentioned in DuPont's patent as a possible raw material to make polyamides, the family of substances to which nylon belongs.

Nylon's most famous application was in the making of stockings that replaced the more expensive silk. With hemlines rising, ladies needed stockings, and the demand for nylon reached staggering proportions. During the Second World War, there were great shortages of stockings as supplies of nylon were diverted to military use in parachutes, ropes and tires. When nylon stockings returned to the consumer market in 1945, there were riots in the streets as women competed for the limited supplies. Nylon, of course, is still with us, and not only in stockings. Tennis racquet strings, windshield scrapers, Velcro, carpets and numerous other items are a testimonial to DuPont's dicey commitment to engage in fundamental research back in the 1920s.

And how did the name nylon come about? One of the early names being considered besides Duparon was Nuron, which implied "new" and also spelled "no run" backwards. Indeed, the newfangled stockings were less likely to run than silk stockings. But Nuron

did not roll off the tongue readily, and a little playing around with
the letters resulted in nylon.

Nike athletic footwear rode to popularity on the basis
of air cushions in the heel of the shoe. The original
"air" in the shoe was actually sulphur hexafluoride,
which has since been replaced by nitrogen. Why
was it replaced?

Sulphur hexafluoride is a greenhouse gas, some 22,200 times
more potent than carbon dioxide, explaining why release of the
gas from old sneakers, which are discarded by the millions annu-
ally, has been connected to climate change. Nitrogen, on the other
hand, makes up about 80 per cent of air and does not present an
environmental problem.

Nike's original air-cushioning system was the brainchild of
Frank Rudy, a former aerospace engineer, who in the early 1980s
came up with the idea of inserting a little air bladder into the heel
of a running shoe. He licensed the idea to Nike, the company
named after the Greek goddess for victory. Nike took the idea and
ran with it, right to Michael Jordan, who was just starting to show
the talent that would eventually make him into the greatest basket-
ball player of all time. The shoe, named Air Jordan, immediately
became popular, but sales really exploded in 1987 with the introduction of a little plastic window in the side of the shoe that made
the air bubble visible to all.

Early versions of the shoe did contain air, but the bubble tended
to deflate as oxygen and nitrogen, the normal components of air,
diffused through the polyurethane. The answer to this problem

was sulphur hexafluoride, an inert gas composed of molecules heavier and larger than those of nitrogen or oxygen. Not only did this gas not diffuse through the polyurethane, but the bubble actually inflated with time as the smaller air molecules diffused into the bubble. This principle can be dramatically demonstrated by blowing up a balloon with sulphur hexafluoride. Instead of deflating, it will actually grow in size as air enters faster than the gas molecules leak out.

Sulphur hexafluoride is indeed a fascinating gas, and it's sometimes used in a classic science demonstration known as the "invisible water effect." An apparently empty fish tank is introduced, and a small aluminum boat is placed inside it. The boat bobs up and down, appearing to float on "invisible" water. As a further conundrum, the invisible water can be bailed out with a cup, with the boat then floating lower and lower in the tank. Of course, there is no such thing as invisible water; the boat is actually floating on the heavier-than-air sulphur hexafluoride. A common finale to the sulphur hexafluoride demonstration is to inhale some. The effect is the opposite of that seen—or, to put it more aptly, *heard*—when helium is inhaled. Instead of speaking like Donald Duck, the voice is dramatically lowered, making one sound more like Darth Vader. This is because the vocal cords vibrate more slowly in the sulphur hexafluoride atmosphere.

Since sulphur hexafluoride is inert, there is no risk of toxicity, at least from the gas. However, there is always the chance of impurities such as hydrogen fluoride being present, and furthermore the gas can displace oxygen from the lungs, making breathing difficult. But the real problem with sulphur hexafluoride is that, after release into the atmosphere, it doesn't break down; it stays around to prevent heat from being dissipated into space. It is this greenhouse effect that stimulated Nike to develop an alternate "green" technology, whereby the air inside the polyurethane bubble really is air—well, almost. It is the major component of air: nitrogen.

The bubble doesn't deflate, because although some nitrogen molecules escape, others diffuse in. Since oxygen also enters, there is actually a small inflation with time.

But the real success of Air Jordan was not due to the gas in the bubble in the heel, but rather to the talented feet inside the shoe.

What do Australian banknotes and disposable underwear have in common?

Both are made of polypropylene. Polypropylene is a plastic with a myriad uses. Like other plastics, it is made of polymers—long molecules made by linking together small molecules, or monomers, in a chain. The monomer used to make polypropylene is propylene, which is produced from natural gas or petroleum. It was first made in 1954 by the Italian chemist Giulio Natta, who probably couldn't have imagined how useful his invention would turn out to be.

Polypropylene can be used to make heat-resistant food containers, carpets, ropes, roofing membranes, laboratory flasks and underwear of several varieties. Since it is warm and has an amazing ability to wick sweat away, it is ideal for thermal underwear. Long underwear made from woven polypropylene fibres is widely used by the U.S. military. But polypropylene can also be extruded into thin sheets used, for example, in diapers and in disposable underwear designed for use during hospital stays or spa visits or by lazy travellers.

And polypropylene sheets can also be turned into money—literally. When forged paper currency began to circulate widely in Australia in the late 1960s, the Reserve Bank of Australia began to search for ways of making banknotes more secure. It took twenty-one years, but polymer chemist David Solomon of the University of Melbourne came up with a method to produce polypropylene banknotes that were essentially impossible to forge. The bills are also extremely durable. This is because they are made of polypropylene that is biaxially oriented—the sheets are stretched in both directions as they are extruded from the hot melt, intertwining the long polymer molecules. If you should happen to forget polypropylene bills in your pocket, they will make the journey through your washing machine unscathed. No surprise, then, that some seventeen countries have followed the Australian example and introduced plastic banknotes, with New Zealand, Romania, Vietnam, Brunei and Papua New Guinea having converted fully. By the way, polypropylene money is completely recyclable. So you may eventually end up wearing it as disposable underwear.

Synthetic fleece, under the brand name Polarfleece, was introduced in the 1980s. What recycled consumer product is it made from?

Beverage bottles made of polyester. Malden Mills, a textile manufacturer in New England, was facing financial problems. Labour costs and taxes were much lower in the southern states, so companies in the northeast were having a hard time surviving. Malden researchers began looking for a product that would distinguish the company from others, and came up with a fleece made of polyester fibres. Fleece, traditionally made of wool, is warm and comfortable, but somewhat heavy. And when it gets wet, it takes a long time to dry. Polyester fleece, on the other hand, is lightweight, water resistant and very warm. Polarfleece, as Malden named its product, was marketed through outdoor outfitter Patagonia and quickly became successful. The popularity of synthetic fleece increased further when Malden Mills' archrival, Wellman Inc., hit on the idea of recycling discarded polyester bottles into chic clothing.

Making clothing from synthetic fibres is not a novel idea. We have already encountered the unsuccessful efforts of Count Chardonnet and the successful efforts of Wallace Carrothers. In the 1980s, as worries about our reliance on a non-renewable resource such as petroleum increased and the environmental movement began to gather steam, plastic recycling took on importance. And used soda bottles made of polyester were ideal. After sorting out non-polyester parts such as caps, the bottles can be put through a sterilizing bath, dried, then crushed into tiny chips that are melted and extruded through spinnerets into fibres that can be turned into fleece.

The energy required to do all this is considerably less than that needed to make fibre from virgin polyester. And each jacket made from recycled polyester keeps about twenty-five bottles out of a landfill. Polyester fleece does not pill-bunch into those annoying little balls and is extremely warm because the pile surface provides space for air pockets that act as excellent insulators. The material is much loved by mountain climbers, but has also been made into underwear for astronauts and, of all things, ear-warmers for calves born in winter. The downside of recycling bottles is the expense of transport. Since the bottles are very lightweight, it takes a large number to make up a ton. To make the process economically viable, sources of used bottles must be found near the factory where they are spun into fibre. And of course hand-sorting recyclable polyester bottles from other plastics is labour-intensive, but machinery is being developed to do this. All in all, being fleeced in this way is a good thing.

ON YOUR METAL

What would an alloy of 92.5 per cent silver and 7.5 per cent copper be called?

Sterling silver. Pure silver is not suitable for jewellery because it is too soft and malleable. Combining it with other metals makes for a much more durable product. Although other metals can be used, copper has proven to be the best companion for silver, improving its hardness without affecting its colour or lustre. Pure silver, also known as fine silver, is sometimes used for intricate jewellery made of thin silver fibres. Silver used to be a popular coinage metal, but the only country that still has silver coins today is Mexico. Compounds of silver are converted to metallic silver upon exposure to light, a process that is the basis of the photographic industry. Indeed, the main use of silver is for the production of photographic film; but with the advent of digital photography, the role of silver is waning.

What metal reacts the most violently with water?

Francium. If you were to drop a piece into water you would see an explosive evolution of hydrogen gas, which would burst violently into flames. A scaled-down version of this phenomenon is seen in the high school chemistry class demonstration involving a piece of sodium. Sodium reacts with water to form hydrogen gas and sodium hydroxide. The metal skitters about the surface of the water and then bursts into flame as the hydrogen being released catches fire.

A piece of potassium performs even more impressively, and it is outdone by rubidium. These metals are all in the same column of the periodic table of elements, meaning they have similar chemical properties. They're called the alkali metals, and all react with water. But the bottom two in the column, cesium and francium, react with almost unbelievable violence. A small piece of cesium tossed into a bathtub filled with water blows the water out of the tub and shatters the tub. If you want to see this fantastic event, search Google Video using the keywords "alkali metal."

Francium would react with even greater vigour, but nobody has ever done the experiment for the simple reason that this element is very rare. It does occur in nature, but it is estimated that there are only a few grams of it in the earth's crust. Francium forms by the radioactive decay of uranium or plutonium, and in turn decays into other elements, so it does not accumulate in the ground. Francium was first isolated in 1939 by Marguerite Catherine Perey, who had been an assistant to Marie Curie and carried on her work after Curie died in 1934. She managed to isolate a small sample of francium as a product of the radioactive decay of actinium. The sample didn't last long—half of it disappeared by radioactive decay every twenty-one minutes—but she did manage to identify the element and patriotically name it after the country where it had been discovered. Too bad there isn't enough francium to be had for tossing into a bathtub full of water. Of course, the experiment would have to be conducted outdoors, but what a sight it would be!

In 2000, the Royal Canadian Mint changed the major metal used to make the Canadian penny. What was the previous metal, and what was it replaced with?

Steel (iron) took the place of zinc. It's a common belief that pennies are made of copper, but that has not been the case since 1996. That was the last year in which the major metal in the penny (to the extent of 98 per cent) was copper. The coins also contained 1.75 per cent tin and 0.25 per cent zinc. Because of the rising cost of copper, beginning in 1997, they were 98.4 per cent zinc, with a copper plating accounting for the other 1.6 per cent. Finally, in 2000, the formulation was changed to 94 per cent steel, 1.5 per cent nickel and 4.5 per cent plating of copper and zinc. A Canadian penny costs about 0.8 cents to produce—a relative bargain, especially when you consider that the U.S. Mint spends 1.4 cents to make each of its pennies. Not a very economical process. The U.S. penny is made of zinc with a thin coating of copper, and the price of zinc has been going up faster than that of gold. I suspect we will one day see the penny disappear as costs continue to escalate. Or perhaps it will be replaced by a plastic coin. But you can imagine the uproar that would generate.

Why would someone put mustard on a coin?

To discover whether or not the coin contains silver. The black substance that deposits on silver as it tarnishes is silver sulphide, which forms upon reaction of silver with traces of hydrogen sulphide. Air always contains some hydrogen sulphide, from volcanic eruptions and petroleum-refining processes, which is the reason that silver left exposed to air will tarnish.

Mustard contains a number of sulphur compounds that slowly release hydrogen sulphide. If a coin is made of silver, coating it with mustard for a few hours will produce a black deposit of silver sulphide. You will have to find an old coin, though: no Canadian coin minted today contains any silver. What *are* they made of? Dimes and quarters contained silver until 1968, when the spiralling price of the metal put an end to its use in coinage. Today, they are made mostly of steel alloyed with a small amount of copper. Both are plated with nickel. The loonie is 91.5 per cent nickel, 8.5 per cent bronze (copper and tin). The toonie is the most interesting coin; the outer ring is made of nickel, while the insert is 92 per cent copper alloyed with 6 per cent aluminum and 2 per cent nickel. Since nickel is attracted to a magnet, the outer ring will be attracted but the insert will not, since such a small concentration of nickel is not enough to cause magnetic attraction. American coins are mostly made of zinc and copper and are not attracted by a magnet. That's one way that vending machines and parking meters distinguish between Canadian and American coins.

When the toonie was first introduced, it presented a challenge to scientifically minded people: how can the insert be removed? Ingenuity came to the fore. Since copper expands more than nickel when heated, it can be expected to shrink more quickly when cooled. Heating a toonie and then plunging it into cold water could cause the insert to fall out. This is no longer the case, since the composition has been adjusted so that the ring and insert expand at the same rate. In any case, it is illegal to experiment in this fashion with coins.

What did the ancient Egyptians refer to as "iron of heaven?"

Meteorites. Iron does not occur on earth in its elemental form. It is found in mineral ores in which it has combined with other elements, such as oxygen or sulphur. To use iron as a metal, it has to be smelted, meaning that the ore has to be heated to a high temperature in the presence of carbon. The carbon strips the oxygen or sulphur from the ore, leaving the elemental iron behind. This process was discovered sometime around 1500 BC by the ancient Hittites of Asia Minor, launching the Iron Age. But iron implements have been found in ancient Egyptian tombs dating back to 3500 BC; in all probability, these were fashioned out of meteorites. Iron is made by the nuclear fusion processes that occur inside of stars, and when a star burns out, remnants in the form of meteorites are scattered. Often these also contain nickel, which prevents the iron from rusting. Ancient Egyptians likely found some of these meteorites and fashioned objects out of the "iron of heaven." These objects were more revered than gold—which is unreactive and *does* occur on earth in its native state—because iron could not be found on earth; it had to come from the heavens.

It is usually easy to blow out candles—except when they happen to be "magic candles." What metal is behind the magic?

Blowing out the candles on a birthday cake is supposed to make your wishes come true. So it's pretty unnerving if they quickly relight of their own accord. How do those "magic candles" sold in novelty stores do it? With magnesium.

There are three conditions for combustion: there must be a fuel that burns, a source of oxygen and some source of heat that allows molecules of the fuel and of the oxygen to become energetic enough to combine with each other. In general, combustion reactions do not occur spontaneously. An unlit candle is always in contact with oxygen from the air but will not light by itself. That is because the oxygen molecules are not banging into the wax molecules with enough energy to cause a reaction.

But when we supply energy in the form of a flame from a match, we melt a little of the wax, convert it to vapour and impart enough energy to the vapourized molecules to react with oxygen. The molecules have been "activated"—they are moving faster, they bang into each other more forcefully and we have a chemical reaction—namely, combustion. This reaction is exothermic—it produces heat and can therefore sustain itself. But if we cool it down by blowing on it, the flame goes out. In the self-lighting candle, small particles of magnesium are incorporated into the wick. This metal gets very hot before it burns. Blowing extinguishes the candle flame because the wax vapour is cooled below its ignition temperature, but the smouldering magnesium stays hot enough to relight the wax after the blowing—that is, the cooling—has stopped. To put the candle out permanently, we have to cool the magnesium by dipping the wick into water or salt.

SCIENCE TO
THE RESCUE

What revolution did Norman Borlaug initiate?

The "Green Revolution," which implemented agricultural technology to prevent famine after the Second World War. The emphasis was on cross-breeding to develop improved, high-yielding strains of wheat, rice and corn, and on the judicious use of fertilizers and pesticides. Without the results obtained from the Green Revolution, the world would not be able to meet basic food requirements. Some estimate that it has saved more than a billion lives.

Once upon a time, not so long ago and in lands not so far away, there was a population explosion. After the Second World War, thanks to improvements in health care, the population in developing countries grew at a rate that outstripped farmers' ability to grow enough food. A terrible hunger crisis loomed. Just as a great threat to global peace had been overcome, imminent famine promised to initiate strife once again. Lord John Boyd Orr, a Nobel laureate and the first director-general of the UN's Food and Agriculture Organization, put it very well: "You can't build peace on empty stomachs." It is this doctrine of "food before peace" that inspired Dr. Norman Ernest Borlaug, recipient of the 1970 Nobel Peace Prize, to initiate efforts to curb hunger. These efforts turned into the phenomenon we have come to know as the Green Revolution.

Born in 1914 in Cresco, Iowa, Dr. Borlaug started his career in forestry before a graduate degree in plant pathology brought him into the world of agronomics. By 1944, Borlaug was in Mexico, acting as director of the wheat program at a research centre funded by the Rockefeller Foundation. Initially, the program was designed to teach Mexican farmers new agricultural techniques, but Dr. Borlaug soon had a more far-sighted research initiative up and running. His first major products were "dwarf" varieties of wheat with short stalks that were able to make use of artificial fertilizer to produce more grain instead of producing tall but nutritionally insignificant stalks. These dwarf varieties are also robust and highly disease resistant, and their growth was not sensitive to the number of daylight hours, making them ideal crops for varied locations across the globe.

The development of the robust dwarf varieties of wheat, and their use in Mexican agriculture, produced extraordinarily high-yielding crops that made Mexico self-supporting in wheat by 1956. No longer did the country have to import the grain; in fact, it was on its way to becoming a major exporter. Once the Mexican agricultural initiative was well established, Dr. Borlaug turned his attention to waning grain production in India and Pakistan.

At first, Borlaug faced resistance to his agricultural innovations in these countries, where people were wary of the new Western technology. They were suspicious of the mutant grains and the agricultural techniques necessary for the high-yield results seen in Mexico. However, dire strains on food reserves, combined with Dr. Borlaug's relentless efforts, soon compelled India, then Pakistan, to give the dwarf wheat a chance. By 1968, Pakistan was completely self-sufficient in wheat and India was well on its way to self-sufficiency as well.

Dr. Borlaug's innovations soon spread to the Philippines and then to parts of Turkey, Afghanistan, Iran, Iraq and many other countries.

In some areas of British Columbia pine forests, little pouches about ten centimetres wide can be seen hanging from the trees. What is their purpose?

To protect the trees from attack by the mountain pine beetle. The little bags contain verbenone, a chemical normally released by beetles as they lay their eggs just under the bark of a pine tree to let other beetles know that "this tree is taken." Using verbenone in this fashion is an attempt to curb the destruction caused by these "bark beetles," as they are commonly called. The destruction is devastating, as is evident to anyone flying over the pine forests of western Canada or the U.S. Rust-coloured pines, the hallmark of beetle infestation, pollute the green landscape everywhere. This pox is not only a visual blight; the sickly brown trees are dry and present a fire hazard. There are also obvious implications for the forestry industry, since any wood harvested from infected trees is of poor quality.

Beetles carry out their dirty work by boring through the bark to lay their eggs, in the process damaging the tubules through which the tree is supplied with water and nutrients. And as if that weren't enough, the beetles also contaminate the tree with the ascomycete fungus, which stains the wood blue and wreaks further havoc on the tubules. Foresters have tried many tactics—thinning forests before an outbreak, removing dead trees, diversifying trees and using various insecticides—but nothing has been able to stem the onslaught.

While insecticides such as carbaryl, bifenthrin or permethrin can dispense with the beetles, they also harm beneficial insects and put animals and people at some risk. Injecting insecticides such as abamectin, emamectin benzoate or imidacloprid into the tree is a safer approach but problematic to apply on a large scale.

There is hope that some advantage may be gained by using the beetles' own chemistry. Once the bugs burrow into the tree, they use some of the tree's chemicals to synthesize verbenone, which is a communicating agent or pheromone. Verbenone can be synthesized in the laboratory and hung on branches in little sachets, protecting the trees from bark beetle attack. While the technique works, its costs make it impractical on a large scale, but it can be used to protect trees near homes, campgrounds or resorts. Given the prospect of global warming, the beetle problem is likely to become worse as larvae are more likely to survive balmy winter temperatures.

Which one of the following lifestyle changes would have the greatest impact on the greenhouse gas emissions attributed to an average North American household: 1) not driving one day a week; 2) eating only locally grown produce; 3) switching to a vegetarian diet; 4) eating only organic foods instead of conventionally produced foods.

It isn't even close. Changing to a vegetarian diet is the most effective way to cut down on emissions. The food consumed by the average North American family is responsible for the emission of almost twice as much greenhouse gas as driving a vehicle.

Although carbon dioxide is the major greenhouse gas, it is by no means the only one. Animals, fertilizers and manure all release methane, which actually traps heat about three hundred times as efficiently as carbon dioxide. Fertilizers also release nitrous oxide, another potent greenhouse gas. Calculating the total amounts of greenhouse gases associated with these activities isn't a simple process. For food, the math has to take into account everything from emissions associated with clearing land and tilling soil to producing fertilizers, transport, refrigeration and cooking. When all of that is taken into account, the results can be expressed in carbon dioxide equivalents.

To the surprise of many, it turns out that the average family is responsible for about eight tons of emissions due to food production, and only about 4.4 tons due to driving. Meat and dairy products account for the biggest chunk of food-related emissions. Raising cattle requires a lot of energy. Producing just one kilogram of beef requires about thirteen kilograms of grain and thirty kilograms of forage. Cows burp and flatulate copiously, releasing lots of methane. Their manure outgases nitrous oxide. Chicken, pork and fish can be produced with a lot less feed, and are therefore responsible for fewer emissions. But eating produce and grain directly, without going through an animal, is the most efficient way to cut down on food-related greenhouse-gas emissions.

A family switching to a vegetarian diet could eliminate roughly four tons of carbon dioxide—equivalent greenhouses gases a year, while giving up driving for one day a week would cut out only about 0.6 tons. Organically raised chickens actually require more energy to produce than their conventionally raised cousins because they are not confined in huge industrial coops. Since they exercise more, they must consume more grain to reach the same weight. And as far as locally grown food versus food transported from afar, there is not much difference. Only about 11 per cent of the greenhouse gases associated with food is due to transport;

83 per cent is linked to the production itself. It is quite clear that as far as greenhouse gas emissions go, vegetarians have it all over carnivores.

How may a move toward novel technologies in the production of chlorine raise the IQ of the general population?

By eliminating the use of mercury, an environmental contaminant that has been linked with reduced IQ. In the classic method of producing chlorine, developed in the nineteenth century, an electric current is passed through a salt solution. Chlorine gas is generated at the positive electrode (the anode), while hydrogen gas and metallic sodium are formed at the negative one (the cathode). This is where mercury enters the picture. The negative electrode is actually a pool of this fascinating metal. Sodium ions from the salt solution are attracted to this electrode, pick up an electron and dissolve in the mercury as sodium atoms.

The mercury, with its load of sodium, then flows into a "secondary cell," where reaction with water converts the sodium into commercially marketable sodium hydroxide. Commonly known as lye, sodium hydroxide is widely used in the petroleum refining, aluminum processing, water treatment, paper production and chemical manufacturing industries. After removal of its sodium content, the mercury is recycled back into the electrolytic cell. But this is where a problem creeps in. The process is not 100 per cent efficient, and some mercury is always lost to the environment, potentially ending up in water systems and contributing to the mercury contamination of fish.

Studies show that consuming mercury-tainted fish during pregnancy or during early childhood can reduce IQ. While adding to the environmental burden, mercury released from "chlor-alkali plants" is by no means the major source of mercury contamination. In the U.S., mercury-cell chlor-alkali plants release roughly three thousand pounds of mercury a year. Compare this to the fifty tons released by electric-power plants that burn coal, which, unfortunately, always has a small mercury content.

Eliminating the production of chlorine is, of course, out of the question. Besides its essential role in water purification, chlorine is widely used to make polyvinyl chloride, or PVC, a commonly used plastic. The pharmaceutical, pesticide and bleach industries also rely on the use of chlorine; it's also essential to the production of titanium dioxide (the classic white pigment in paints). While chlorine production cannot be eliminated, the use of mercury can be. Alternate processes have been developed. The most efficient one, known as the membrane cell process, uses a Teflon-like membrane to separate the chlorine from the sodium ions during electrolysis. Governments around the world are moving to introduce legislation to force chlor-alkali plants to shift to the new technology or close down. According to current plans, plants that use the mercury process will be eliminated within a decade. By then, someone will probably have discovered that the fluorinated polymer used in the membrane cell process presents some sort of environmental problem. But hopefully it will be a smaller problem than we now face with mercury.

What would be the purpose of piping chilled ammonium carbonate into the chimney of a power plant that produces electricity by burning coal?

To reduce carbon dioxide emissions, and hence reduce global warming. Roughly 50 per cent of all electricity in North America comes from burning coal. The heat produced by the combustion process is used to convert water into steam, which is then used to turn a turbine to generate electricity. Carbon dioxide emissions from such power plants make up about 30 per cent of all such emissions. The result is the much discussed, and feared, greenhouse effect.

One way of curbing the global warming attributed to this effect is to capture carbon dioxide before it is released into the atmosphere. Various technologies are being explored, one of which uses chilled ammonium carbonate. When this substance is combined with carbon dioxide and water vapour in flue gas, it forms ammonium bicarbonate, a solid that can be recovered from the chimney. When the recovered ammonium bicarbonate is heated, it reverts to ammonium carbonate and carbon dioxide. The carbonate can be recycled back into the chimney, and the carbon dioxide can be concentrated under pressure into a liquid. This has commercial value for freezing foods and for putting the fizz into soft drinks. But far more carbon dioxide would be captured by this process than could be sold. So the idea is to sequester the excess underground—by pumping it down to a depth of at least eight hundred metres, where the atmospheric pressure is great enough to ensure that the carbon dioxide remains in a liquid state.

Experiments have shown that carbon dioxide can indeed be stored in underground geological formations. In fact, if there are oil or gas deposits nearby, the carbon dioxide can even be used to help pump these fuels out of the ground before settling in the empty wells. Admittedly, the technology is not without its critics. Some say that the gas can leak out either slowly or, in the case of underground seismic activity, massively. And being heavier than air, carbon dioxide can concentrate in surface hollows, where it can asphyxiate animals or people. Right now, though, the biggest concern is cost. The process used to remove carbon dioxide from power plant emissions would jack up the price of electricity. So when it comes to reducing this greenhouse gas, people will have to open their wallets as well as their mouths. But the ammonium carbonate technology does go to show that, when push comes to shove, human ingenuity can rise to the surface. And in this case, it can come up with a method to sink carbon dioxide emissions.

What common feature links the following substances: potassium acetate, sodium chloride, calcium chloride, magnesium chloride and urea?

They are all used to melt snow and ice on roads in winter. All of these substances interfere with the formation of ice crystals and can be used to melt ice. They do, however, differ in effectiveness, potential harm to the environment, and cost.

Sodium chloride, or common salt, is cheap and can melt ice at temperatures as low as −20°C. But it can also damage soil and vegetation, contaminate surface and ground water and speed up the corrosion of concrete and metals. Corrosion of metals, a process whereby the metals react with oxygen, requires the transfer of electrons among substances, and such transfer is facilitated by the presence of ions, such as sodium and chloride. Substances that dissolve to form ions in solution are called electrolytes and speed up the rusting process. That's why cars in Canada rust and those in Arizona do not. It's also why airplanes not in use are stored in the Arizona desert.

Not all electrolytes speed up corrosion to the same extent. Potassium acetate is much more environmentally friendly than salt but is twenty times as expensive. Calcium chloride melts snow and ice much faster than sodium chloride, and is less corrosive, but damages vegetation and wildlife. It also costs more than salt. Magnesium chloride is also less corrosive, but costs five times as much as salt and stops working once the temperature drops to −15°C. Urea is non-corrosive and doesn't damage vegetation, but only melts ice down to −4°C.

There is yet another issue with salt. It works more effectively if the grains can be prevented from clumping. To reduce caking, small amounts of sodium ferrocyanide are added. In the presence of sunlight, this can break down and release cyanide, which can be washed into waterways and damage aquatic life.

Ethlylene glycol or propylene glycol are the substances used to de-ice airplanes because they are non-corrosive. These liquids are collected and recycled, but some is inevitably lost to the environment. Obviously, there is no perfect way to melt ice and snow. But not using these substances would result in loss of life. As with so many scientific issues, it is a question of evaluating risk versus benefit.

Europe is planning to ban HFC-134a, the chemical currently used in automobile air-conditioning systems, because of its contribution to global warming. What are automobile makers planning to replace it with?

Carbon dioxide. That may sound strange, because carbon dioxide is the main greenhouse gas in terms of volume released into the atmosphere. But there are many other gases that, molecule per molecule, have a greater warming effect on the earth than carbon dioxide. That effect is measured in terms of the global warming potential, or GWP. By definition, the GWP of carbon dioxide is 1. HFC-134a, the refrigerant currently used in cars, has a GWP of

1,300, and will be disallowed after 2011, when Europe introduces legislation to ban refrigerants with a GWP over 150.

When HFC-134a was introduced as a replacement for ozone-depleting freons in the 1980s, there was no concern about global warming. Back then, the issue was excess exposure to ultraviolet light due to a depletion of the protective ozone layer in the strato-sphere. HFC-134a solved that problem, but introduced another one. When it inevitably escaped from refrigerators and air condi-tioners, it contributed to global warming. Hence the need for a replacement for the replacement. Carbon dioxide fits the bill because of its lower global warming potential and its inertness toward ozone. Of course, it also has to work as a refrigerant. And it does. It can be compressed into a liquid, and when the pressure is reduced the liquid changes back into a gas, drawing heat from the surroundings. This is the essence of the workings of any refrig-erator or air conditioner. Carbon dioxide is already readily avail-able because it is produced in huge amounts for the carbonation of beverages and for fire extinguishers. The problem is that greater pressure is needed to liquefy it than is needed for HFC-134a, so automobile air-conditioning systems have to be completely rede-signed. That's why automobile makers are looking at alternatives that would not require such a major change.

Animal fat is being used by several companies as a source of renewable energy. What is produced from the fat?

Biodiesel. Any type of fat can be converted into diesel fuel by reacting it with methanol or ethanol. Most biodiesel production relies on vegetable oils, but ConocoPhillips, a major American oil company, believes that animal fat can be converted into biodiesel economically and has signed a deal with Tyson Foods, the world's biggest meat producer, to supply it with raw material.

Animal fat that is collected after slaughter is usually turned into ingredients for soap, cosmetics or pet food, but conversion to biodiesel may be a more attractive proposition given that tax breaks will be available. While corn and soy can be used to produce biodiesel, these crops can also furnish food for people. By contrast, making biodiesel from animal fat makes use of a byproduct of meat production. No extra animals will be slaughtered for their fat. ConocoPhillips believes that it can produce about 3 per cent of all its diesel from rendered animal fat.

One of the environmental benefits of diesel derived from animals is that it contains no sulphur, eliminating the production of sulphur dioxide. Once the biodiesel has been produced, there is no way to tell where it came from—there will be no steak or bacon aroma coming out of the tailpipe. Of course, there are ethical issues involved, since people who are opposed to the slaughter of animals would be unable to tell whether the diesel fuel they purchase includes a component sourced from slaughterhouse byproducts. The meat industry is certainly not an environmentally friendly one. According to some estimates, it is responsible for more global warming emissions than all the cars, trucks and planes in the world combined. So at least by using the fat left behind after slaughter, the meat industry is making some amends, since biodiesel produces fewer greenhouse emissions than regular diesel.

Hydrogen is being touted as the fuel of the future. How is it produced commercially?

Not, as is commonly thought, by passing an electric current through water. While this can be done in the laboratory, the method is not viable on a commercial scale. Most hydrogen is produced by combining natural gas with steam. This means that we are still using a non-renewable resource, namely natural gas, to produce hydrogen. If solar, tidal or wind power could be used to supply the energy needed to break water down into hydrogen and oxygen, then we would be talking about a possible solution to the energy crisis. Furthermore, when methane is reacted with steam to produce hydrogen, the other byproduct is carbon monoxide. This can be made to react with more water to produce more hydrogen, but there is then another byproduct, namely carbon dioxide. So even hydrogen production has a greenhouse component. The solution is to pass the carbon dioxide that is formed into a solution of sodium hydroxide, which converts it to sodium carbonate and keeps it from getting into the atmosphere. There is no free lunch.

Rainwater is always acidic to some degree because of three gases that dissolve in water to produce acids. What are these three gases?

Carbon dioxide, sulphur dioxide and nitrogen dioxide. There is a common misconception that all acid rain is due to human activity. This is not the case. Carbon dioxide is always present in the air, and it dissolves in water to form carbonic acid. Sulphur dioxide is released by volcanic activity and forms sulphuric acid, while nitrogen dioxide forms when the heat of lightning allows nitrogen and oxygen in the air to combine. Nitrogen dioxide then dissolves in water to form nitric acid. All of these reactions are "natural."

But most of the acidity in rainwater is caused by human activity. Coal always contains small amounts of sulphur, and the burning of coal to produce electricity spews huge amounts of sulphur dioxide into the air. Cars are a major culprit in terms of nitrogen dioxide production, since their engines produce enough heat to allow nitrogen and oxygen to combine.

Acid rain has numerous environmental effects. Acidification of soil can impair the growth of some crops and trees, and aquatic life can be adversely affected. Buildings made of limestone—also known as calcium carbonate—also erode when exposed to acid rain. Many measures have been introduced to control acid rain. Techniques have been developed to convert sulphur dioxide as it passes up chimneys into sulphates, which can then be used to produce commercially useful gypsum. Recirculating some of a car's hot exhaust gases into the intake manifold allows the engine to run at a cooler temperature and release less sulphur dioxide. In some cases, excess acidity in aquatic systems has been neutralized by the addition of a base such as limestone.

HISTORY
DEPARTMENT

February 2009 marked the 200th anniversary of the birth of Charles Darwin and the 150th anniversary of the publication of *On the Origin of Species*. What was Darwin's nickname in school?

"Gas." It seems that the great naturalist's first interest in school was chemistry, and his interest in the subject gave rise to the name. But let's let Darwin tell the story in his own words:

> Towards the close of my school life, my brother worked hard at chemistry, and made a fair laboratory with proper apparatus in the tool-house in the garden, and I was allowed to aid him as a servant in most of his experiments. He made all the gases and many compounds, and I read with great care several books on chemistry, such as Henry and Parkes' *Chemical Catechism*. The subject interested me greatly, and we often used to go on working till rather late at night. This was the best part of my education at school, for it showed me practically the meaning of experimental science. The fact that we worked at chemistry somehow

got known at school, and as it was an unprecedented fact, I was nicknamed "Gas."

I was also once publicly rebuked by the head-master, Dr. Butler, for thus wasting my time on such useless subjects; and he called me very unjustly a "pococurante," and as I did not understand what he meant, it seemed to me a fearful reproach.

Well, Darwin was right. Pococurante was not a compliment; it means "caring little." Hardly an apt description of the founder of evolutionary biology. Since Charles was not good at what the head-master thought were useful subjects, such as learning the verses of Virgil or Homer, he was taken out of school by his father and sent at an early age to medical school in Edinburgh. The only entrance requirement here was money, which Charles's father, a prominent physician, had plenty of.

Darwin didn't take to medicine, recalling in his autobiography that "the education at Edinburgh was altogether by lectures, and these were intolerably dull, with the exception of those on chemistry by Hope." The brutality of surgery in those days before anaesthesia also left a mark on Darwin. On two occasions he rushed out of the operating theatre, unable to cope with the suffering he witnessed. Basic science, he decided, was much more of an attraction. In 1831, he took a post as a naturalist on HMS *Beagle*, which was setting off on a voyage of scientific exploration. On the Galapagos Islands, Darwin was struck by the presence of fourteen species of finches unique to those islands. Pondering the question of how so many species could be found in such a small area got him thinking about natural selection. In 1859, Darwin published *On the Origin of Species*, which is the foundation of modern biology. But it all started with an interest in chemistry.

Light-coloured straw hats were very popular in the early twentieth century. What chemical was used on the usually darker-coloured straw to achieve this effect?

Hydrogen peroxide. Today, hydrogen peroxide has numerous uses, but one of its first commercial uses was to bleach the straw used for making hats. The discovery of this important chemical can be traced back to Napoleon's desire to develop batteries. The emperor had been greatly impressed by a demonstration he witnessed, performed by the Italian scientist Alessandro Volta. Volta's battery consisted of discs of copper and zinc separated by discs of paper or cardboard soaked in salt water. Attached to the top and bottom of this "voltaic pile" was a copper wire; when Volta closed the circuit, electricity flowed through the pile. Napoleon was so taken by this demonstration that he made Volta a count for his discovery. He also urged French scientists to try to improve on Volta's method. It was this stimulus that resulted in the discovery of hydrogen peroxide by Louis-Jacques Thénard in 1818.

Thénard was investigating the potential of various metals as components in batteries when he found that exposing barium oxide to oxygen resulted in the formation of barium peroxide—which, when dissolved in hydrochloric acid, yielded hydrogen peroxide. He called the resulting solution "oxygenated water." Early use of this oxygenated water was in the bleaching of textiles, and the straw used to make hats. Today, the chemical finds widespread use as an antiseptic, as an ingredient in hair-colouring products and as a treatment for polluted water. The drugstore version, a 3 per cent solution, can be used as an effective disinfectant. It is sold in brown plastic bottles for two reasons. Small amounts of metal ions, such as sodium or calcium, which can leach out from glass containers, can hasten the decomposition of hydrogen peroxide to water and oxygen. Ultraviolet light can also trigger this decomposition. Brown plastic bottles avoid both problems.

Who is the first person known to have been successfully treated for eczema by being given a preparation containing live bacteria?

Adolf Hitler. Probiotics, or micro-organisms that confer a health benefit, are popular nutritional supplements these days, with abundant claims being made about reducing gastrointestinal problems, controlling the proliferation of disease-causing microbes and alleviating allergies. Yogurt containing live bacteria flies off the shelves, and capsules containing various probiotic cultures crowd the shelves in health food stores.

The idea that certain bacteria can promote health was introduced by Russian bacteriologist Ilya Mechnikov, who believed that longevity was linked to the presence of lactic acid–producing bacteria in the gut. Later, other bacteria were linked with health benefits, including a strain of E. coli that was originally isolated by German army surgeon Dr. Alfred Nissle in 1917 from the feces of a soldier who did not develop diarrhea during a severe outbreak of infection by shigella bacteria. A commercial product known as Mutaflor was developed based on Nissle's strain of E. coli and was administered to Hitler by his personal physician, Theodor Morell, when the Führer complained of suffering from a skin rash and intestinal gas. Apparently it worked, giving Hitler confidence in the physician's abilities.

Morell was on the fringes of medicine; he eventually treated Hitler with a cacophony of drugs that included hormones, opiates, stimulants and enzymes, along with vitamins and "natural" preparations containing animal tissues such as placenta, cardiac muscle, liver and bull testicles. But with his probiotics, Morell may have struck the right note. There is an increasing amount of evidence that such bacterial cultures can produce benefits. One theory is that, in recent years, the variety of bacteria in our gut has changed due to our extreme cleanliness and a decline in breast-feeding, resulting in a reduction of worthy targets for our immune system. A consequence is that the immune system begins to attack invaders such as pollen that don't actually represent a risk.

In one recent double-blind, randomized study, subjects suffering from hay fever were given a daily dose of live *Lactobacillus casei Shirota* in a milk beverage, and their immune status was monitored over five months. Those given the live bacteria had lower levels of IgE antibodies, which suggests a reduced risk of symptoms when exposed to an allergen. A number of trials have also examined the effect of supplementing the diet of babies with probiotics, with ambivalent results. A few have shown a reduction in the propensity for allergies, but most have not. Results with digestive problems are more encouraging. While there is a great deal of unsubstantiated hype about probiotics, there is also an increasing amount of science testifying to possible effectiveness for some conditions. But researchers have to nail down which bacteria may be useful, and for what condition. It isn't a case of "one size fits all." Morell also treated Hitler with strychnine for his digestive problems, but unfortunately did not use a large enough dose.

What McGill University graduate said he hoped his tombstone would say only "He brought medical students into the wards for bedside teaching?"

William Osler. Osler's wish never came true, for the simple reason that he has no tombstone. His ashes rest in an urn in the Osler Library of Medicine at McGill University, perhaps the most appropriate final resting place for the man who has been described as the father of modern medicine.

Osler was an excellent clinician, but it was his revolutionizing of the teaching of medicine that brought him everlasting fame. He graduated with a medical degree from McGill in 1872, returning to the university as a professor in 1874. After ten years at McGill, Osler moved to the University of Pennsylvania as chair of clinical medicine, and then in 1889 to Johns Hopkins University, where he became the first chief of staff. He finished his career at Oxford University, succumbing to the Spanish flu during the 1919 epidemic at the age of seventy.

Osler believed that medical students spent too much time listening to theoretical lectures and insisted that his students get to patients' bedsides early in their training. He established the practice of clinical rotations for third- and fourth-year medical students and introduced the idea of residency, whereby newly graduated physicians start their careers in hospitals under a pyramid system, learning from more senior doctors and teaching junior ones. Osler was a great collector of books on medical history, which he left to McGill and which now form the core of the Osler Library of the History of Medicine, the most extensive collection of such references in Canada.

In a famous speech entitled "Aequanimitas," delivered when he left the University of Pennsylvania, Osler put forth his views on the practice of medicine. A physician should always maintain an outward expression of calmness and coolness, even under difficult circumstances. Empathy is important, he maintained, but the objective practice of medicine requires a degree of detachment from patients. Physicians can't fix everything and should calmly accept whatever life has to offer. Osler had a humorous side, although not everyone saw the humour, especially when he gave a talk suggesting that when a man's productive life is over, at age sixty-seven, after a year of contemplating his accomplishments, he should be put to permanent rest with chloroform. Osler also wrote mischievous letters to medical journals under the pseudonym Egerton Yorrick Davis. One of his letters, uncritically published by *The Philadelphia Medical News*, reported the supposed phenomenon of "penis captivus," in which it becomes impossible to withdraw the penis after sexual activity.

In 1783, a strange object fell from the sky in Gonesse, France, and was promptly attacked by the villagers with pitchforks. What was the object?

The first-ever hydrogen-filled balloon. The dream of flying goes back to Greek mythology and the famous story of Icarus, who made wings of feathers and wax but crashed when he flew too close to the sun and the heat melted the wax. One of the first recorded cases of the ups and downs of science!

Man's first successful flight was not with wings, but with a balloon filled with hot air. The idea of using hot air for flight came to Joseph Montgolfier in 1782 as he supposedly watched laundry drying over a fire, sometimes billowing upwards. He concluded that smoke was responsible, and put the idea to a test by building a little pyramid-shaped wooden framework covered with silk and holding it above a fire. It quickly ascended. Within months, Joseph and brother Étienne held a public demonstration of a large hot-air balloon made of paper and cloth. Joseph's idea was that a balloon large enough to carry men could be built and put to military use.

Jacques Charles, a Parisian nobleman, heard about the Montgolfiers' discovery and became extremely interested. He had already developed a fondness for studying gases and had even organized public lectures on their behaviour for the elite of the capital, who could afford to pay the required fee. Charles often demonstrated the properties of "inflammable air," as hydrogen was called in those days. Henry Cavendish had shown in 1766 that this lighter-than-air gas could be generated by reacting acids with iron, a process that Charles often featured in his lectures. When he heard about the Montgolfiers' balloon adventure, he assumed they had used a hydrogen-filled balloon, since this was the only lighter-than-air gas he knew about.

Charles decided to outdo the Montgolfiers and build a bigger balloon, one that could perhaps carry humans. To cover the cost of making the huge amount of hydrogen that would be needed, tickets were sold to the public to view the hydrochloric acid being poured into large drums filled with iron filings. It must have been quite a spectacle indeed.

Just three months after the first Montgolfier demonstration, Charles launched his hydrogen balloon in Paris, to the delight of an enthusiastic crowd. The balloon landed less than an hour later in the little village of Gonesse, where it caused massive panic. Perhaps fearing the appearance of some monster from the sky, the villagers attacked the object with scythes and pitchforks.

Jacques Charles would eventually build a hydrogen balloon capable of carrying a human—in fact, himself. On December 1, 1783, along with fellow balloononaut Marie-Noël Robert, Charles flew for two hours, covering a distance of some thirty miles. He had lost out, though, to the Montgolfiers for the distinction of being the first human to fly in a balloon. In November of the same year, Pilâtre de Rozier and François Laurent had taken to the air in a Montgolfier-designed hot-air balloon and flown for twenty-five minutes.

The brothers themselves never flew and never accepted Charles's explanation that it was hot air, and not smoke, that made their balloon rise. But Joseph Montgolfier's idea of using balloons for military purposes did come to fruition. In 1794, the French used a tethered balloon to get a vantage point over the Austrians, resulting in a French victory at the Battle of Fleurus. By the time the American Civil War rolled around, ballooning had become established and Professor Thaddeus Lowe was able to convince President Abraham Lincoln of its potential in battle. In 1861, at the battle of Bull Run, Lowe was able to observe Confederate positions and direct cannon fire at the surprised enemy.

Bakelite was the first plastic made out of a totally synthetic polymer. How did Leo Baekeland finance the research that led to this invention?

With his first invention, Velox photographic paper. Leo Baekeland was born in Belgium in 1863 and as a youngster developed an interest in photography, which led to an infatuation with chemistry. He became thoroughly enthralled with the process that allowed pictures to be displayed on paper. Baekeland experimented with making his own film, but was short of the cash needed to purchase silver nitrate, the key component. However, he did have a silver watch that a relative had given him. Baekeland cleverly reacted the silver with nitric acid and produced his own silver nitrate.

After obtaining a degree in chemistry, Baekeland was offered a fellowship to travel overseas to further develop his interest in the chemistry of photography. In the U.S. he found a research position with the E. & H.T. Anthony Company, a leading manufacturer of photographic paper. There, he began the work that led to the invention of Velox. At the time, strong sunlight was needed to develop photographs—an obvious limitation for photographers. But Baekeland's new paper could be developed under artificial light, which meant kerosene lamps.

George Eastman, who had already founded the Kodak company, heard of Baekeland's discovery and invited him to Rochester to discuss buying the rights to the invention. Musing about how much to ask for, Baekeland decided on $25,000, a huge amount at the time. Eastman, however, took him by surprise. He opened the conversation with "How does $750,000 sound for Velox?" To Baekeland, of course, it sounded fine. The stunning sum fixed him financially for life. Now he could devote his time to whatever pleased him. And what pleased him was the search for some sort of insulating material to replace shellac.

The electrical industry was burgeoning, and shellac, an ideal insulator, was in short supply. Baekeland knew that, years before, Adolf von Baeyer had combined phenol and formaldehyde to produce a black guck that, to him, was useless in his search for novel dyes. But, Baekeland thought, it might work as an insulator. It took about three years to come up with a method of reacting these chemicals to produce a synthetic alternative to shellac. The secret was a type of pressure cooker, appropriately named the Bakelizer, which was used to allow the molecules of phenol and formaldehyde to join together to make long molecular chains, or polymers.

Bakelite, as Baekeland modestly named the novel substance, could do far more than serve as an insulator. When the freshly made hot fluid was poured into moulds, it would upon cooling take on the shape of the mould permanently. It was a thermosetting plastic, meaning it could not be melted again. Soon, radio housings, telephones, rosary beads, billiard balls, jewellery, distributor caps, propellers and even coffins were produced from Bakelite. The age of plastics was on the way, and by 1980, the annual manufacture of plastic overtook that of steel. The family friend who, in the 1967 hit movie *The Graduate*, whispers into newly graduated Ben's ear, was right on the money: "I just want to say one word to you. Just one word: plastics." And it all started with Bakelite.

Why does a barber's pole have a red stripe?

It represents blood. During the Middle Ages, monks were required to shave the crowns of their heads, a function commonly performed by itinerant barbers. Also, under ecclesiastic law, monks had to be periodically bled. This was supposedly a symbol of piousness, of devotion to God. Barbers began to attend to this duty as well. They would travel with a "flag" of a white cloth dipped in blood to indicate that they would attend to anyone who needed to be bled. This early mode of advertisement eventually was transformed into the barber's pole. And the pole began to symbolize more than haircuts and bleeding.

Barbers began to expand their role and became quasi surgeons, specializing in sewing up wounds and extracting teeth. They also had a sideline in whitening teeth, by dabbing them with nitric acid. This did produce an immediate whitening but destroyed the teeth in the long run by wearing away the protective enamel. But at least one sixteenth-century barber-surgeon, Ambroise Paré, made an important contribution to medicine.

Barbers in those days worked under the guidance of physicians, who thought themselves above menial jobs like cutting and scalding. Why scalding? Because physicians thought that gunpowder was poisonous, and gunshot wounds therefore had to be treated with boiling oil to destroy the poison. Unfortunately, if the bullet didn't kill the victim, the scalding often did. During the siege of Turin in 1537, Paré ran out of oil and for some reason substituted a cold mixture of egg yolks, oil of roses and turpentine. To his surprise, the soldiers treated with this mixture fared better than those who had been scalded. And thus ended the brutal practice of pouring hot oil into bullet wounds.

The French-trained Paré was a religious sort and thought he had had divine assistance in making his observation. That's why he introduced the oft-repeated phrase, "I dressed the wound, but God healed him." With the advent of modern science, surgeon-barbers have, of course, disappeared. In fact, even barbers are disappearing. They have been replaced by hairstylists, who fiddle about with the hair for what seems like hours, and you can't even amuse yourself by watching the turning of the barber pole.

In the late 1940s, long-playing records made of polyvinyl chloride (PVC) made their debut. What material did PVC replace in record manufacture?

Shellac. This resinous secretion of the female *Laccifer lacca* beetle has had a number of other uses since it became commercially available in the 1800s. Shellac can be moulded under heat and pressure and was used to make jewellery boxes, combs and even early dentures.

The lac beetle is native to Southeast Asia, but it is only the female that is useful, at least in terms of providing shellac. I guess, though, there would be ño females without the males, so they do serve a purpose. The female has an unexciting life, entirely spent attached to a tree. She sucks sap from the bark and converts it to a resin, which she uses to glue herself to the tree. This is where the male finds her, performs his reproductive duties and leaves the female encased in what becomes her resinous tomb. Indian peasants discovered that the resinous material could be scraped from trees, melted, filtered from insect parts and used to provide a lustrous preservative coating to wood.

The production of shellac was not easy; a single pound was the result of some fifteen thousand lac beetles toiling away for six months. As long as shellac was used as a wood coating, the demand could be met. After all, labour in India was cheap. But with the advent of the age of electricity, demand increased sharply because shellac turned out to be the best electrical insulator available. All of a sudden the price soared, and demand increased further when shellac turned out to be a better material than hard rubber for making phonograph records. It wasn't ideal, though, given that records made of shellac were quite fragile and could only hold a few minutes of music per side.

Then along came polyvinyl chloride, which allowed records to have narrower grooves and be played back at lower speeds with less surface noise than shellac. In 1948, Columbia Records began marketing the long-playing vinyl record developed by Peter Goldmark, and shellac was shunted aside, at least as far as the recording industry was concerned. Today, shellac is still sometimes used as a wood preservative and, interestingly, in the food industry. Since it is perfectly edible, it is used as a coating on candies and is sprayed onto fruits to provide a thin coating to prevent moisture loss.

Which twentieth-century agricultural scientist was famously criticized by *The New York Times* for referring to God as "Mr. Creator" and denying evolution?

George Washington Carver, better known as the Peanut Man. He was an agricultural researcher best known for finding multiple uses for peanuts. Carver also dedicated his life to teaching former slaves such as himself agricultural techniques so that they could become self-sufficient, and even built a mobile school for this purpose.

Carver was a devout Christian who credited God for guiding his research. He did not believe in evolution and took everything in the Bible as literal truth. He endured a great deal of criticism from the scientific community and the press, including *The New York Times*, whose November 20, 1924, article entitled "Men of Science Never Talk That Way" admonished him for his research methodology. Carver's unwavering faith probably came from the fact that he had had an extremely difficult childhood and found solace in prayers and the company of fellow Christians. George was born into slavery in 1864 and, while still an infant, he, his mother and his sister were kidnapped from their owner by the Confederate army and sold in Arkansas, a common practice.

The family's original owner, a German-American immigrant by the name of Moses Carver, hired a professional to find them, but only George was found alive, orphaned and near death from whooping cough. Because of this misadventure, George suffered from bouts of respiratory disease and was too weak to work in the field. Instead, he spent his days wandering the field and getting to know the plants. He became so knowledgeable that he came to be known as the "Plant Doctor" by Moses Carver's neighbours. Because of his failing health, George was never expected to live beyond his twenty-first birthday, but he went on to live a long and successful life until the age of seventy-eight. With age, his faith grew as well.

George Washington Carver also received some good press. In 1941, *Time* magazine dubbed him the "Black Leonardo," a comparison to Leonardo da Vinci. His numerous accomplishments included not only agricultural advances, but also mentoring children, improving social and racial relations, painting, poetry, advocacy of sustainable agriculture and appreciation of plants and nature. Upon returning home one day, George took a bad fall down the stairs and died of complications on January 5, 1943. On his gravestone was written the simplest and most meaningful summary of his life: "He could have added fortune to fame, but caring for neither, he found happiness and honor in being helpful to the world."

In 1915, Dr. Cluny Macpherson, a McGill graduate, impregnated a piece of canvas with a solution of sodium thiosulphate and sodium carbonate. This led to what invention that would be widely used during the First World War?

The gas mask. On April 22, 1915, the Germans released chlorine gas over a four-mile-long front, causing about fifteen thousand British casualties. The British retaliated some six months later, and by the time the carnage was over, about 1.3 million casualties due to gas warfare had piled up. Dr. Macpherson came up with a simple gas mask that could neutralize chlorine. The mixture of sodium thiosulphate and sodium chloride reduced chlorine gas to sodium chloride, which is harmless. As new gases were developed, gas mask technology had to keep up. When phosgene was introduced, phenol and sodium hydroxide were used in masks to neutralize it.

Chloropicrin required a canister filled with charcoal to adsorb most of the gas, and soda-lime and permanganate to destroy the rest.

What did the ancient Egyptians use natron for?

Making mummies. The Egyptians believed in immortality, and central to this belief was the idea that the soul left the body at death, only to return at some later time to reclaim and resurrect the remains. Obviously, during the waiting period, the body, which housed the soul, had to be preserved somehow.

Unfortunately, the process of putrefaction starts soon after death. As oxygen distribution by the blood grinds to a halt and the immune system ceases to be active, anaerobic micro-organisms begin to attack and break down the tissues. Soon an overpowering smell emerges as proteins decompose to yield a variety of odoriferous compounds with such delightful names as cadaverine and putrescine. The latter substance, also known as ptomaine, is slightly poisonous and is found in the decaying flesh of all animals.

Like any other form of life, the microbes responsible for decay require water to survive. The ancient Egyptians undoubtedly became aware of this when they noted that corpses buried in the dry sand of the desert didn't decompose. A little experimentation revealed that natron, a naturally occurring mixture of sodium carbonate, bicarbonate, sulphate and chloride found along the shores of lakes in the vicinity of Cairo, worked even better than sand at dehydrating a body. So the mummification process was born.

The internal organs, where putrefaction was known to start, were removed and the body cavities washed with a disinfecting solution of wine and spices. The brain was removed through the nostrils and discarded—apparently there would be no need for it in the afterlife. The heart, which was the seat of the soul, was left in the body, but the other organs were separately pickled in jars, supposedly to be reassembled at the time of resurrection. The whole body was then packed in natron for at least forty days, after which it was wrapped in fabric and finally dipped into a gummy resin. The mummy was finished!

But as far as we know, only in the movies have mummies come back to life. Has the failure of mummies to sit up and walk out of their tombs put an end to the mummy business? Certainly not. A U.S. organization known as Summum is offering high-tech mummification. Chemical preservatives, along with fibreglass and polyethylene wraps, may be just what yuppies are looking for to ensure that their bodies, immaculately sculptured by years of aerobic exercise, are not going to be defiled by those nasty anaerobic microbes.

Incidentally, there is no truth to the rumour that when railroads were first built in Egypt, the steam engines were fuelled by a plentiful supply of mummies. On the other hand, stories about ground-up mummies being used as medicine in Europe are based in fact. Of course, the medication was totally useless, except if the patient truly believed in it. Otherwise, it was as effective as skull scrapings from executed criminals, which were also marketed as a cure-all.

What industry in the nineteenth and early twentieth century was associated with the destruction of the jawbones of its workers?

The match industry. The condition was known as phossy jaw and was caused by exposure to white phosphorus. When strike-anywhere matches were first produced in 1833, they seemed like a godsend. Unfortunately, they ended up sending some of their makers to god. Lucifers, as these early matches were called, were made by dipping wooden sticks into a mixture of white phosphorus, potassium nitrate (saltpetre) and glue. Friction heated up the phosphorus, which then ignited and burned with the help of oxygen released by the potassium nitrate.

Dipping the sticks into the mix, known as "compo," exposed the workers to the vapours of phosphorous, which commonly infiltrated the jaw through tooth cavities—which, of course, were common at the time. The gums would swell, teeth would hurt and abscesses would begin to form in the jawbone, which would then become disfigured. A fetid discharge oozed from the jaw with an odour so powerful that it made its victims social pariahs. The only solution at the time was surgical removal of the bone, which then allowed patients to survive, albeit miserably. Sometimes so much phosphorus built up in the bone that samples removed by surgeons actually glowed in the dark.

Eventually, in the early 1900s, white phosphorus matches were banned everywhere, but by this time numerous workers had suffered from phossy jaw. And they need not have. As early as 1855, matches made with safe red phosphorus were known, but lucifers continued to be made, thanks to industrial greed.

We may be seeing a modern version of phossy jaw emerge as a consequence of using drugs known as bisphosphonates, which are widely prescribed to treat osteoporosis by preventing bone loss. The phosphorus in these drugs can cause a rare condition called osteonecrosis, or bone death, in the jaw.

What is believed to be the first synthetic pesticide?

Lime sulphur. The use of this substance dates back to the 1840s, when it was used to control a type of mildew on grapevines that had been exported from the U.S. to France. The disease was devastating to the French wine industry, reducing production by 80 per cent. Lime sulphur turned out to be an effective way to control the problem. It is prepared by boiling lime, or calcium hydroxide, with sulphur. The result is a substance that can be described chemically as calcium polysulphide but is commonly known as lime sulphur. The reddish-yellow water solution of this substance can be sprayed on trees and plants to combat fungi, bacteria and insects. Solutions are extremely alkaline and can cause skin damage and terrible eye injuries. If ingested, it reacts with stomach acid to yield highly toxic hydrogen sulphide, the stuff that gives rotten eggs their fragrance.

Lime sulphur is made from sulphur, but sulphur itself can be used as a pesticide. In fact, it is probably the oldest effective insecticide. The Romans burned "brimstone" as a fumigant, relying on the toxicity of the sulphur dioxide released. They apparently also used some other natural insecticides, such as the gall from a green lizard to protect apples from worms and rot. Sulphur can be used without burning as well. Sulphur dusts are toxic to various mites and are effective against powdery mildews. Since sulphur occurs in nature, it is acceptable in organic agriculture. So is lime sulphur, because its ingredients both occur in nature. Isn't it curious that the first synthetic pesticide ever introduced is now regarded as a staple of organic agriculture? Where does one draw the line between natural and synthetic? Well, all synthetic pesticides are made from substances found in nature. Maybe there are more chemical reactions involved than just heating lime and sulphur, but does that really matter?

Why did Napoleon tell Josephine, "Don't wash. I am coming home!"

Apparently, the Little Emperor was enticed by Josephine's sweaty aroma. Armpit fragrance has actually been scientifically examined—most notably by Dr. George Preti of the Monell Chemical Senses Center in Philadelphia. He had volunteers wear absorbent pads under their arms and then analyzed the contents. Preti was able to isolate some forty compounds, but one in particular turned out to be responsible for sweaty-armpit smell: 3-methyl-2-hexenoic acid, which apparently forms when bacteria attack a protein released by the sweat glands. Why Napoleon would have been attracted by this smell is uncertain, although other components of underarm secretions have been linked with aphrodisiac effects. Androstenone, which in the amounts produced in sweat has no fragrance, is one of these candidates. It actually is known to be the sex attractant of the wild boar, and it just may have some effect in humans. In one widely quoted study, seats in a theatre sprayed with the compound were more likely to be occupied than other seats. Interesting, but the study has never been reproduced. Could it be that, with all our devotion to a smell-free existence, we are undermining our love life?

What gas derives its name from the Greek word meaning "lazy" or "inactive?"

Argon, from the Greek *argos*. Argon is indeed inactive; it does not undergo any chemical reactions. The gas was discovered in 1894 and became the first element to be labelled as "noble" because of its lack of reactivity. Argon was later joined by helium, neon, krypton, xenon and radon in the family of inert gases.

The discovery of argon was a result of some ingenious collaboration between the English physicist Lord Rayleigh and the Scottish chemist William Ramsay. Rayleigh had prepared a sample of nitrogen by decomposing ammonia but noticed that the density of the gas was slightly different from a sample of nitrogen he had isolated from air. At the time, air was known to contain nitrogen, oxygen, carbon dioxide and water vapour. Rayleigh had cleverly removed the other gases and found that the gas that remained, supposedly pure nitrogen, weighed slightly more than the other sample of pure nitrogen he had made from ammonia. He consulted Ramsay about this conundrum, and the two concluded that the nitrogen sample from air must contain a small amount of another gas, one heavier than nitrogen. This gas turned out to be argon, the third most abundant gas in the atmosphere. Both Ramsay and Rayleigh were awarded Nobel Prizes for the discovery in 1904— Ramsay in Chemistry and Rayleigh in Physics.

Commercially, argon can be produced by distilling liquid air. The main use of the gas is based upon its chemical inertness: it's used— instead of air—to fill incandescent light bulbs because it will not react with the tungsten filament. It is also used in fluorescent bulbs and in welding operations where air needs to be excluded. When aluminum is welded with an electric arc, for example, an atmosphere of argon prevents the formation of aluminum oxide that would interfere with the welding process. Curiously, the wealthy eccentric scientist Henry Cavendish had stumbled upon argon in 1785 but did not recognize his discovery. Cavendish subjected a sample of air to various chemical reactions and removed the products. No matter what he did, he was always left with about 1 per cent of the original volume of air that would not react chemically.

Before electricity, streets were lit by gas lamps. Where did this gas come from?

Burning coal. When coal is burned it produces a mixture of gases composed of hydrogen, methane, carbon monoxide and various hydrocarbons. Historically, of course, people have used various means to battle darkness. The first idea, traced back to about 70,000 BC, was to take a hollow rock, shell or other such natural object, fill it with moss soaked in animal fat and ignite it, probably by rubbing two sticks together.

The lighting of public streets was first attempted in 1417, when Sir Henry Barton, mayor of London, ordained "lanterns with lights to be hanged out on the winter evenings between Hallowtide and Candlemasse." Coal gas came along near the end of the eighteenth century when William Murdoch, a Scottish engineer, recognized its potential as an alternative to lamps that used oil or tallow as fuel. By 1794, Murdoch was producing coal gas from a small retort containing heated coals, attached to a three- or four-foot iron tube through which he piped the gas before sending it through an old gun barrel and igniting it to produce light. Murdoch spent the next ten years experimenting with and improving that technology, determining the best temperature to heat coal to obtain the maximum quantity of gas, and finding ways, such as the use of lime, to remove objectionable smells.

William Fairbairn, himself a Scottish engineer, recorded that Murdoch occasionally used his gas as a portable lantern. "It was a dark winter's night, and how to reach the house over such bad roads was a question not easily solved. Mr Murdoch, however, fruitful in resource, went to the gasworks where he filled a bladder which he had with him, and, placing it under his arm like a bagpipe, he discharged through the stem of an old tobacco pipe a stream of gas which enabled us to walk in safety to Medlock Bank."

Unfortunately, despite his pioneering work with gas, Murdoch never made any money from his invention for the simple reason that he never patented it. Gas lighting has by now been relegated to the pages of history, but if you feel nostalgic, you can still go to the Park Estate in Nottingham, which still uses gas lighting to retain its historical character.

What did the Swedish astronomer Anders Celsius invent?

Not the thermometer. That was the brainchild of a contemporary of Celsius, a German physicist by the name of Daniel Fahrenheit, who invented the alcohol thermometer in 1709 and the mercury thermometer in 1714. Fahrenheit's scale was a bizarre one, and to this day his reasoning is not completely understood. He knew that the freezing point of water could be reduced by dissolving salt in it, and he defined zero as the temperature of a water-ice-salt mixture. Then, for reasons known only to himself, he defined body temperature as 96 degrees and divided the scale between zero and 96 into equal units. By this measure, water boiled at 212 degrees, so this became the upper limit. Fahrenheit's reckoning of body temperature missed the mark—it's actually 98.6. Confusing, right?

But Celsius came to the rescue. He defined zero as the freezing point of pure water and 100 as the boiling point of pure water. Many thermometers still feature both scales, but mercury thermometers for home use are being replaced by alcohol thermometers or electronic ones. That's because the environmental pollution and health risks associated with broken thermometers can be considerable.

Thomas Edison's wife, Mina, preferred jewellery made of amber. Why was this particularly appropriate?

Because Edison's name is intimately associated with electricity, and the ancient Greek word for amber is *elektron*.

Any account of the history of electricity should begin with an observation made more than 2,500 years ago. When the ancient Greeks began to navigate the Black Sea, they opened up trade routes to the Balkans, where they became acquainted with the fossilized resin of pine trees. The substance seemed well suited for the making of jewellery, and it proved to be readily formulated into rings, bracelets and earrings. Then one day, some fastidious Greek lady or gentleman, in the process of cleaning amber jewellery with a piece of fur, noted that it developed an ability to attract threads, feathers or bits of fluff. At the same time, the materials that were attracted to the amber repelled each other.

The phenomenon remained no more than a curiosity until Elizabethan times, when the English physician William Gilbert discovered that substances other than amber also had such properties and coined the term "electric" to describe their behaviour. The attractive force the Greeks had observed so long ago came to be known as electricity. Now the experiments came fast and furious. When two amber rods were rubbed, or electrified, with fur, they repelled each other. So did two glass rods. But the amber rods attracted the glass rods, at least until they touched. Then they lost all electrical power!

Today, we can explain this behaviour. When amber is rubbed with fur, electrons are transferred from the fur to the amber, giving the amber a negative charge; the fur, meanwhile, which is now electron deficient, assumes a positive charge. Like charges repel, while unlike charges attract each other. Thomas Edison's electric bulb glowed as electrons flowed through a heated-up filament. So it is appropriate that amber was Mina Edison's favourite gemstone.

MEDICINE CABINET OF CURIOSITIES

Why are pharmaceutical companies that market sustained-release forms of the painkiller oxycodone carrying out research to make pills that cannot be crushed?

Oxycodone is a semi-synthetic opiate painkiller that is widely prescribed for severe pain. It was first synthesized in Germany in 1916, when pharmaceutical companies were in a race to develop morphine analogues that would retain the drug's painkilling effect but eliminate its addictive potential. Efforts centred around isolating compounds from opium—the white exudates collected from poppies—and altering them chemically in the laboratory, hence the expression "semi-synthetic."

Heroin was the first of these, but it did not perform as "heroically" as had been hoped. Then came the conversion of another opium component, thebaine, into a compound that was named oxycodone on account of its chemical similarity to codeine, another opium ingredient. It turned out to be an effective painkiller without producing the same kind of rush as morphine or heroin, and therefore had somewhat less potential for abuse. But in high doses, especially if mainlined, oxycodone was capable of producing euphoria and causing addiction with the usual withdrawal symptoms of nausea, insomnia, extreme anxiety and muscle pain.

Prescriptions for oxycodone increased significantly with the development of a sustained-release form known as Oxycontin, which was capable of controlling pain for periods up to twelve hours. But the popularity of Oxycontin also led to extensive diversion of the drug to the illicit market. Addicts took to crushing the tablets and either snorting or mainlining the resulting powder for a quick rush. To try to prevent abuse, pharmaceutical companies are developing pills that cannot be crushed. One idea is to blend the active ingredient with gelatin to formulate a rubbery substance that can be deformed when hit with a hammer but cannot be crushed. Furthermore, the oxycodone cannot be extracted by soaking in alcohol or water. If the U.S. Food and Drug Administration approves the new formulation, it will go on the market as Remoxy, a painkiller that will hopefully defy abuse.

Alpharma, a New Jersey pharmaceutical company, is working on a different approach. It is working on an opiate-containing pill that would have an inner core made of naltrexone, a drug that counters the effects of opiates by binding to opiate receptors. The idea is to coat the core with a material that prevents the naltrexone from being released if the pill is swallowed, but releases its content if the pill is crushed. But with the novel formulation, even if the opiate enters the bloodstream, it will not produce a high because of the antagonistic effect of naltrexone. Undoubtedly, though, ingenious addicts are already researching ways to circumvent these attempts at tamper-proof pills.

An American veteran suffering from Gulf War syndrome was prescribed Obecalp. When he did a little research into this drug, he became upset. Why?

Obecalp is placebo spelled backwards. The veteran had been given a sugar pill that contained no active ingredient and assumed that the physician had not taken his condition seriously. Mike Woods, like a number of other veterans, suffered from Gulf War syndrome, a collection of symptoms including numbness in the extremities, chest pain and occasional blackouts. There is no doubt that veterans suffer from it, but the question is what exactly they are suffering from. All sorts of theories have been put forward, ranging from exposure to chemical warfare agents to aspartame-sweetened soft drinks that released methanol when subjected to the heat of the desert. But many physicians believe that the symptoms originate in the mind, and they therefore feel it is the mind that needs to be treated. And that is why Woods's physician prescribed Obecalp, a sugar pill that is stocked by some pharmacies for use by physicians who feel that a placebo is appropriate treatment.

Ethical questions certainly have been raised about giving an inactive pill to a patient without his or her consent. But the argument can be made that if a patient is suffering from a condition for which there is no known treatment, there is nothing to lose, and maybe something to gain, by prescribing a placebo. After all, placebos are known to offer relief in many conditions, with success rates often approaching 50 per cent. And actually, whether a placebo really contains no active ingredient is debatable. It may not contain a physiologically active substance, but it can certainly trigger physiological activity.

This was recently shown in an elegant fashion by researchers at Columbia University who placed volunteers in a PET scanner to investigate brain activity as their arms were subjected to uncomfortable heat after having been treated with a topical cream. When the volunteers were told that the cream was a powerful painkiller, activity in parts of the brain that are normally stimulated by opiates was significantly increased, while no such effect was noted when they were told that the cream had no medicinal value.

Obviously, then, the relief afforded by a placebo can be very real, and prescribing a sugar pill is not immoral, as some have suggested. Of course, that's assuming there is no known proven other treatment. Unfortunately, in the case of Mike Woods, the placebo did not work, which led to his feeling of having been somehow tricked because the physician did not consider his condition to be real. He was wrong on that count: the physician considered the condition very real but, having no known treatment to offer, decided to take a chance on a placebo. This is no worse than trying a variety of different medications when a physician is frustrated, a practice that is a common feature of medicine.

What poison played a critical role in the development of the blood pressure medication captopril?

The venom of the Brazilian pit viper, *Bothrops jararaca*. In 1898, two Swedish researchers made an interesting discovery. They injected an extract of the kidney of a dog into another dog and found that the animal's blood pressure soared. The active ingredient in this extract turned out to be an enzyme called renin, which, together with another enzyme secreted by the lungs, produced angiotensin II, a powerful constrictor of blood vessels. It was the enzyme secreted by the lungs, known as angiotensin converting enzyme (ACE), that came to be a major player in the control of blood pressure.

Drugs that block the action of this enzyme, the so-called ACE inhibitors, have become standard therapy not only for high blood pressure, but for congestive heart failure and heart attacks as well. The key discovery that led to the production of captopril, the first ACE inhibitor, was made in 1965 by British pharmacologist John Vane and his post-doctoral student, Sergio Ferreira from Brazil. Ferreira had written his doctoral thesis on the effects of the venom of the poisonous Brazilian pit viper and noted that it caused severe intestinal contractions. He determined that this was due to the presence of an enzyme-inhibiting factor. Ferreira then sought to work with Vane because the latter was carrying out research on enzyme inhibition, although his work had nothing to do with intestinal contractions. Rather, Vane was studying the enzymes involved in blood pressure control.

Similarities between the chemical action that caused intestinal contractions and high blood pressure suggested that the snake venom could lead to a reduction in blood pressure by inhibiting the action of angiotensin converting enzyme. And it did, at least in the test tube. Injecting snake venom into humans was not a viable option, but Vane was a consultant for the pharmaceutical company Squibb and suggested that its researchers study the chemical structure of the venom and design molecules that might retain its blood pressure–lowering activity without the toxicity.

Squibb researchers David Cushman and Miguel Ondetti did just that. They identified the part of the venom molecule that lowered blood pressure and synthesized a variety of compounds that included the active fragment. One of these became the blockbuster captopril, which went on the market in 1981 as Capoten. It was soon joined by Merck's ACE inhibitor Vasotec, which became the company's first billion-dollar drug. As with any drug, ACE inhibitors can cause side effects; coughing is probably the most common one.

Understanding how ACE inhibitors led to a lowering of blood pressure stimulated another idea: instead of preventing the formation of angiotensin II, why not try to block its activity? That research eventually resulted in the angiotensin II receptor agonists like Novartis's Diovan (valsartan), which by 2003 was generating two and a half billion dollars in sales every year. And all because of snake venom.

What did the Rolling Stones refer to as "Mother's Little Helper?"

Valium. The 1967 song poked a little fun at diazepam, the drug that had been introduced in 1963 as Valium and quickly became a best-selling anti-anxiety medication. In "Mother's Little Helper," the yellow pills helped a housewife cope with her bratty kids and difficult husband. And it wasn't only housewives who were getting a little chemical help from diazepam. Comedian Milton Berle quipped that the definition of a Valium addict was a patient who takes more Valium than his doctor.

Jacqueline Susann's famous 1966 novel *Valley of the Dolls* told the tale of ambitious young women coping with life in New York City by resorting to "dolls," a euphemism for Valium. By the late 1970s, two and a half billion Valium tablets a year were helping Americans calm down, but worrisome signs of addiction were already emerging. Barbara Gordon's personal account in *I'm Dancing as Fast as I Can*, published in 1979, signalled the end of the Valium craze as she described how quitting the medication cold turkey had landed her in an insane asylum. Valium is still around today, although its chemical descendants, such as Xanax and Ativan, are more likely to be prescribed.

As is so often the case in science, the discovery of Valium was a happy accident. Leo Sternbach, at the Hoffmann-La Roche research labs in New Jersey, had become interested in developing a drug to compete with chlorpromazine, the first effective antipsychotic drug, which had recently been introduced by a rival company. Looking at the structure of the molecule, Sternbach was reminded of some compounds he had synthesized in a quest for dyes some twenty years earlier as a post-doctoral student in Krakow. Now he began to tinker with the molecular structure of some of these compounds to see if they could be converted into drugs that had an activity similar to chlorpromazine.

One by one the new compounds were tested but failed to produce results. Sternbach's superior put a halt to the research and the samples were discarded, save for a couple that had not been tested. These stayed on a laboratory shelf for two years until Earl Reeder, another researcher, decided to tidy up. He asked Sternbach what they were, and the two decided to test them just to finish off the work that had been started, fully expecting to see the usual negative results. But to their surprise, one of them turned out to have a sedative effect, and after it was shown to tame lions and tigers at the San Diego Zoo, it was introduced into medicine as Librium. A slightly altered molecule, with less of a bitter flavour and greater potency, was introduced three years later as Valium, the name deriving from the Latin *valere*, for "being strong." And it *was* strong, earning La Roche six hundred million dollars a year at its peak. Although Valium had critics because of its addictive potential, Dr. Sternbach was convinced that his invention had prevented suicides and saved marriages. He himself, however, did not pop his famous little pill. He claimed it made him feel depressed.

The first one was made with collodion. Today octyl-2-cyanoacrylate is used. What is it?

A transparent bandage for cuts and burns. Collodion is a solution of nitrocellulose in alcohol and ether, first formulated by Louis Ménard in 1846. When spread on the skin, the solvents evaporate and leave behind a transparent plastic layer. Its use petered out when adhesive bandages such as Band-Aids were developed. There was also a problem with flexibility; the nitrocellulose film often developed cracks. Now the concept has been reborn, and flexible, transparent plastic bandages are available, thanks to octyl-2-cyanoacrylate and spider silk. That may sound like a strange connection, but were it not for an attempt to find a substitute for spider silk, we would not now have transparent bandages. And that's not all we wouldn't have—we wouldn't have super glue either!

During World War II, threads made from spider webs were used to make the crosshairs on gunsights. The black widow spider's web was especially suitable. Still, researchers searched for a more available and more standardized source of fine hairs. That's when Dr. Harry Coover of Kodak Labs found that polymerizing chemicals called cyanoacrylates into polycyanoacrylates yielded a plastic that could be drawn into thin threads. It didn't work out for use in gunsights—or for airplane canopies, which was another Coover idea—but it did eventually find a use. Coover had noted the sticky nature of the material and tried it as a glue. Now, *that* worked! The discovery also gave Dr. Coover a chance to appear on television. *I've Got a Secret* was a popular show in the late 1950s. Coover appeared on the show with his secret, which was a super glue used to stick two metal cylinders together and hoist the host, Garry Moore, into the air.

When the Vietnam War came along, polycyanoacrylates were pressed into service on the battlefield. When used as a spray, they could glue tissues together and prevent blood loss in emergency situations. The stuff used at the time was methyl-2-cyanoacrylate. It worked well but sometimes broke down, releasing formaldehyde and cyanoacetate, both of which are irritants.

A flurry of research followed, and eventually octyl-2-cyanoacrylate replaced the methyl derivative because it turned out to be far less irritating. This is the material that was eventually approved as Dermabond or Liquid Bandage. The interesting aspect of the chemistry here is that the final product actually forms on the skin. The bottle contains the monomer octyl-2-cyanoacrylate, which, when exposed to moisture in the air, reacts to form polyoctyl-2-cyanoacrylate. And all of this because of a need for crosshairs on gunsights.

Alka-Seltzer is often taken by people to relieve headache and simultaneously neutralize excess stomach acidity, yet the product contains citric acid. What is the role of the citric acid?

Alka-Seltzer contains Aspirin, sodium bicarbonate and citric acid. The citric acid reacts with the bicarbonate to produce the fizz, which does little more than impress people that something is happening. Sodium bicarbonate is better known as baking soda because it can cause baked goods to rise as it liberates carbon dioxide gas upon reaction with an acid. In Alka-Seltzer, some of the baking soda reacts with citric acid to produce the fizz, but since there is an excess of bicarbonate there is some left over to neutralize excess stomach acidity. The Aspirin, of course, is effective against headaches so that Alka-Seltzer is a useful medication for someone simultaneously suffering from an upset stomach and a headache, as can happen after a particularly pleasurable night out. To witness the efficiency of gas production, just take an Alka-Seltzer tablet, place it in a film canister with a little water and step back. You'll see the power of carbon dioxide as the lid is blown off in a spectacular fashion.

What natural substance was the first local anaesthetic to be introduced into medicine?

Cocaine. Credit for the discovery of cocaine as a local anaesthetic is usually attributed to Dr. Karl Koller, an Austrian opthamologist who, in 1884, demonstrated this property by dropping a solution of the drug into the eyes of frogs and guinea pigs. He then went on to experiment on some of his colleagues and on himself, clearly proving that cocaine drops effectively desensitized the human eye.

While Koller was the first to use cocaine as an anaesthetic in eye surgery, he was not the first to note the local anaesthetic effect of the compound, which occurs naturally in the leaves of the South American coca plant. That honour actually goes to Friedrich Wöhler, the German chemist who is regarded as the father of modern organic chemistry. Wöhler had garnered scientific fame by making urea, a compound found in human urine, from ingredients that did not come from living sources. With this single experiment he destroyed the notion that substances found in living systems, which at the time were referred to as "organic," could not be reproduced in the laboratory because they contained some sort of "vital force."

Years later, Albert Niemann, working in Wöhler's lab at the University of Göttingen, isolated the active principle of the coca shrub, which Wöhler then named cocaine. He noted that when applied to the tongue, cocaine had a numbing effect, but he never exploited the discovery. The idea of using cocaine as an anaesthetic was actually Sigmund Freud's. The famed Viennese physician, who would come to be known as the father of psychoanalysis, had a notion that it could be used in the treatment of morphine addiction. During the course of this investigation, Freud found that rubbing cocaine on the skin led to a loss of sensation and brought this to the attention of colleague Karl Koller, who went on to use the drug successfully in eye surgery.

Freud would likely have shared the credit for discovering the local anaesthetic effect of cocaine had he not left Vienna to pursue his future wife, Martha Bernays. This chase apparently was also a result of Freud's experimenting with cocaine and his discovery that ingesting a small dose of the drug increased his libido. Freud excitedly wrote to his fiancée, "Woe to you, my princess, when I come. I will kiss you quite hard and feed you until you are plump. And if you willfully resist, you shall see who is stronger, a gentle little girl who doesn't eat enough or a big wild man with cocaine in his body." History does not record the result of this contest, or whether Freud needed cocaine to sweep Martha off her feet. Today, cocaine is still sometimes used in ear, nose or throat surgery, but it has largely been replaced by other local anaesthetics—such as procainamide—that are synthetics modelled on the molecular structure of cocaine.

Introduced to it by Sir Walter Raleigh, Queen Elizabeth I of England remarked, "I don't like this herb." What herb was she referring to?

Tobacco. Raleigh reputedly gave the queen a pipe to smoke, which made her so sick that she believed she had been poisoned. Raleigh was one of Elizabeth's favourite courtesans, having gained the monarch's favour for his role in fighting Irish rebels. However, when he married one of the queen's ladies in waiting without her permission, he was imprisoned in the Tower of London. Raleigh, though, had enough money to buy his way out, and he established a colony in North America which he named Virginia to flatter Elizabeth, the "virgin queen."

Interestingly, Raleigh himself never set foot in Virginia but did import the tobacco and potato plants that grew there. In fact, he became the first British potato planter *and* the first smoker, using a long pipe. Tobacco is believed to have been first grown in America as early as 6000 BC, and Native Americans long used the plant both in religious and medical practices. In fact, when Columbus came to America in 1492 he was offered a gift of dried tobacco leaves. Soon, sailors introduced tobacco growing to Europe, where the plant served mostly as a drug, supposedly able to cure everything from bad breath to cancer.

But it seems that Raleigh really did play a major role in promoting the smoking of the leaves. This seemed so unusual at first that one day while smoking his pipe, one of Raleigh's servants thought he was on fire and doused him with a bucket of water! It wasn't long before some of the dangers of tobacco were being realized, and by 1610 Sir Francis Bacon had noted that quitting the smoking habit was very difficult.

Raleigh later explored South America, looking for the legendary El Dorado, the city of gold, but never found it, for the simple reason that it did not exist. He did find something else, though. He learned that the natives used a type of poison, extracted from a plant, on their arrows. This was curare, a drug that would turn out to have a medical use in controlling muscle contractions during surgery.

Raleigh was not a favourite of King James I, who succeeded Elizabeth, and he was eventually imprisoned again on trumped-up charges, including not having brought back the promised gold from America. He was later condemned to death and was beheaded in 1618. When he saw the axe that would be used, he mused, "This is sharp medicine, but it is a physician for all diseases and miseries." Raleigh's head was embalmed and presented to his wife. The condition of his lungs is unknown. But we do know that since that time tobacco has caused a great deal of misery. Indeed, Queen Elizabeth knew nothing about the risk of smoking, but she was probably the first person to be involved in the acrimonious debate about the habit that persists to this day.

ILLICIT
SUBSTANCES

Diego Maradona shared the honour of being named footballer of the twentieth century with the great Pelé. Maradona's off-field activities were, however, less than honourable. In 1994, he was booted off Argentina's World Cup team. Why?

He tested positive for ephedrine, a banned substance. Ephedrine occurs naturally in a variety of plants of the species Ephedra, and has a close molecular similarity to adrenalin, the "fight or flight" hormone produced by the body's adrenal glands. Like adrenalin, ephedrine is a stimulant, raising blood pressure and dilating the bronchial tubes. Derived from the herb ma huang, ephedrine has long been used, particularly in traditional Chinese medicine, to treat asthma. Since it constricts blood vessels, ephedrine has also been used as a decongestant and as a drug to raise dangerously low blood pressure.

Ephedrine's appetite-suppressing effect popularized its use as a weight-control drug until evidence accumulated about its potential adverse effects on the heart. Baltimore Orioles pitcher Steve Bechler's death in 2003, apparently after taking an ephedrine-containing supplement to lose some weight, stimulated authorities to ban the use of the substance in dietary supplements. Bechler's widow sued the manufacturer of the particular supplement involved in her husband's death, eventually settling for one million dollars.

Unfortunately, the drug also lends itself to other modes of abuse. Students claim that it can increase the ability to concentrate, and athletes use ephedrine for extra energy. This is not surprising, given ephedra's chemical similarity to amphetamine. It was the quest for improved performance that got Maradona into trouble at the 1994 World Cup. He tested positive for ephedrine but claimed he had consumed the drug unwittingly when he drank an American version of the energy drink Rip Fuel, which, unlike the Argentinian drink, contained ephedrine.

Later, Maradona revised the story and claimed that he had a prior agreement with the Football Association to use ephedrine for weight loss so that he could perform up to the American public's expectation at the World Cup being held in the U.S. Ephedrine was not the only drug in Maradona's life; he has a long history of alcohol and cocaine abuse. In 2004, Maradona suffered a major heart attack after overdosing on cocaine. After retiring, Maradona fought a constant battle with weight, eventually undergoing stomach bypass surgery. By 2008 he appeared to have turned his life around, and to the surprise of almost everyone was named manager of the Argentinian national team.

Why are drug manufacturers switching from pseudo-ephedrine to phenylephrine in their decongestant formulas?

Unlike pseudoephedrine, phenylephrine cannot be used by clandestine chemists to synthesize the illegal stimulant methamphetamine. Nasal congestion is the result of swelling of blood vessels in the nose, and since pseudoephedrine is an effective vasoconstrictor, it can effectively relieve the symptoms of congestion. But pseudoephedrine is not specific to constricting blood vessels in the nose; consequently, it can lead to increased blood pressure.

Although the compound occurs naturally along with ephedrine in the ephedra bush, commercial production is based on using a special yeast to ferment a mixture of dextrose and benzaldehyde. As early as 1892, Japanese and Chinese scientists isolated ephedrine from the ephedra bush and realized it was the active principle in the traditional Chinese asthma remedy known as ma huang, which was an extract of the bush. By the 1920s, ephedrine in its pure form had become a popular asthma remedy in North America, but production stumbled when a civil war in China resulted in a severe shortage of ma huang.

Researchers looked for a substitute and found it in the related compound amphetamine, which had first been synthesized by the Romanian chemist Lazar Edeleanu in 1887. There had been no use for this substance until the ephedrine shortage, when Gordon Alles at the University of California, Los Angeles found it to be a useful substitute for ephedrine in the treatment of asthma. He also noted something else: amphetamine was a powerful stimulant! Consumers soon discovered this effect as well, reporting "a sense of well being and a feeling of exhilaration" as well as "lessened fatigue in reaction to work."

These effects were so powerful that, during the Second World War, the Allies, Germans and Japanese all freely distributed amphetamine to their forces to keep them alert. After the war, amphetamine abuse became so widespread that the drug's use had to be severely curtailed, limiting it essentially to the treatment of attention hyperactivity disorder. This is when clandestine chemists swung into action to meet the demand for artificial exhilaration.

Underground labs that produced amphetamine sprung up, but synthesis proved to be difficult. However, a closely related compound, methamphetamine, first synthesized from ephedrine in Japan by Akiro Ogata in 1918, proved an easier target. Although ephedrine was a controlled substance, pseudoephedrine was a common ingredient in cold remedies and was readily converted into methamphetamine. Production of this illegal substance, known on the street as meth, speed, ice or crank, became so widespread that the availability of pseudoephedrine had to be curbed. One way of achieving this was by replacing pseudoephedrine in cold remedies with phenylephrine. Although phenylephrine isn't as effective a decongestant as pseudoephedrine, it cannot be converted into meth.

In 1898, the German pharmaceutical company Bayer introduced a drug that derived its name from the fact that it made people feel unusually powerful and courageous. What was this drug?

Heroin. Opium, an exudate of a species of poppy, is perhaps the oldest known painkiller. Its major active ingredient is morphine, which is wonderful for pain relief but highly addictive. After morphine was isolated from opium in the 1800s, chemists began to wonder if a slight alteration in the structure of the molecule might result in a drug that retained the painkilling effect but was no longer addictive. Researchers at Bayer found that treating morphine with acetic acid yielded diacetylmorphine, which did not appear to be addictive and furthermore made patients feel "heroic." They therefore named the drug heroin and introduced it as a non-addictive version of morphine. Heroin was also sold as a cough remedy and was commonly advertised on the same billboard as Bayer's leading drug, Aspirin. It didn't take long, though, to discover that heroin was indeed addictive and was not a solution to the morphine problem. By 1924, the Heroin Act in the U.S. had made the manufacture and sale of heroin illegal. Unfortunately, the synthesis of heroin from morphine is not difficult, and there is plenty of illegally manufactured heroin around to bring misery to multitudes.

Many farmers in Kansas have installed surveillance cameras to keep an eye on the tanks of ammonia they use as fertilizer. Why?

Ammonia is one of the ingredients needed to make "crystal meth," the stimulant illegally sold on the street. Criminals make the stuff in clandestine labs from substances available in over-the-counter decongestants. Ammonia is one of the reagents needed in the synthesis, and hoodlums have been stealing ammonia from farmers, necessitating the use of surveillance cameras.

THE
BEAUTY
PAGES

Galen, the second-century Greek physician, blended together mineral oil, beeswax and rose-water to make what?

Moisturizing cream. The essential ingredient in all moisturizing creams is some sort of oily substance that acts as a barrier through which water cannot evaporate. It may be termed a moisturizer, an emollient or a lubricant—interchangeable terms, for all intents and purposes. Both mineral oil and beeswax are effective moisture barriers, but they feel greasy. In order to cut the oily feel, Galen added rosewater. There is no doubt that the cream was effective, but it left a lot to be desired in terms of texture and appearance. No amount of mixing could keep the ingredients from separating into different layers.

Modern cosmetic chemistry, however, has solved this problem through the use of emulsifiers, chemicals that allow immiscible substances to combine to form a homogeneous product. An everyday example of this phenomenon is the use of lecithin, found in egg yolk, to prevent vinegar and oil from separating in a salad dressing. So modern creams have not really improved on the

effectiveness of Galen's original concoction, but the use of emulsifiers has ensured a tempting, smooth consistency.

Today's emollients can be chosen from a variety of waxes,
vegetable and mineral oils or silicones. Of course, moisturizing
products also contain some water. This has nothing to do with
moisturizing effectiveness; the water content just determines
whether the final product will be a cream or a lotion. To blend
the ingredients together, emulsifiers with such tongue-twisting
names as propylene glycol stearate are incorporated. Then there
may also be fragrances, preservatives and thickeners.

All creams and lotions contain some variation on these basic ingredients. Are they then all the same? As far as moisturizing goes, the
answer is yes. But how is it that people have such distinct preferences?
Since skin types and skin dryness differ, some people may prefer a
heavier moisturizer; others, a lighter one. Fragrance can also be a determining factor. So can price. The general assumption is often that a
more expensive product must be better. But many a study has shown
that, when trials of effectiveness are carried out in which the subjects
are unaware of which product they are using, there are just as many
favourable comments for the cheap as for the expensive products.

In the 1920s, hair removal by what method—widely
advertised as "absolutely painless and needle free"—
became a medical disaster?

X-rays. In 1905, Wilhelm Röntgen made one of the most famous
medical discoveries of all time when he noticed that a beam of electrons passing through a partially evacuated tube gave off invisible
rays that caused fluorescent materials to glow. Intrigued by this

observation, he put his hand in front of a fluorescent screen and noted that an image of his bones formed. Apparently, bones blocked the novel mysterious rays, which were christened x-rays.

It wasn't long before physicians were using x-rays to peer into the human body. But they did note a side effect: x-rays made hair fall out. That gave Austrian physician Leopold Freund an idea: why not use x-rays to get rid of unwanted body hair? It certainly seemed to be an improvement over tweezing or using strong alkalis. Early reports of x-ray hair removal were very positive. Even a bearded lady was said to have been cured of her affliction. When some women in France complained that the treatment had made them sick, Freund retorted that this was to be expected from the French, who had a "hysterical character."

American physician Albert Geyser, who had experimented with x-rays, jumped on the bandwagon despite being painfully aware of the risks. He had lost a couple of fingers on his left hand due to x-ray burns. But this had stimulated him to create an x-ray tube that he claimed delivered "softer" rays. Soon the Tricho Sales Corporation was born, and clinics using Geyser's technology began to sprout up around the U.S. and Canada. Women flocked to the clinics and revelled in their new, hairless skin. But not for long.

Soon there were complaints of wrinkled skin and lesions where the supposedly safe x-rays struck, and by 1929 the evidence for damage was so clear that the *Journal of the American Medical Association* published an article noting that women, "in their endeavour to remove a minor blemish, have incurred a major injury." Nobody really knows how many cancers were triggered by exposure to depilatory x-rays, but it's a good bet that thousands of women suffered the consequences of what turned out to be a reckless application of a very useful technology.

What contribution can roosters make to women's beauty. How?

The rooster's comb is a flap of skin that swells in response to testosterone. When a rooster's comb is erect, the rest of him is ready for action. The material that fills in the comb to cause swelling is known as hyaluronan, previously called hyaluronic acid. Each hyaluronan molecule is a chain of small sugar units joined together. The substance can be extracted from the comb, and a purified hyaluronan gel has been approved as an injectable material to fill in facial wrinkles. The company that manufactures it is certainly crowing about the results.

What cosmetic processes involve the application of thioglycolic acid followed by the application of hydrogen peroxide?

Permanent waving of hair and the straightening of curly hair. Hair is essentially a network of protein molecules that are in turn composed of chains of amino acids. Adjacent protein molecules are held together by bonds known as disulphide linkages. Essentially, two sulphur atoms link the protein chains, sort of like rungs on a ladder. To change the structure of the hair, these sulphur-sulphur bonds have to be ruptured so that the protein chains can move relative to each other. The most common chemical used to accomplish this is thioglycolic acid. Once the protein chains have been disengaged, the hair can be shaped on curlers or combed until it is straight. Hydrogen peroxide is then used to reform the sulphur-sulphur linkages and hold the hair in its new shape. Permanent curls, of course, grow out as new hair is formed.

DOMESTIC
SCIENCE

The list of ingredients reads: sodium tallowate, glycerine, titanium dioxide and tetrasodium EDTA. What is it?

Soap. The making of soap is one of the oldest of all chemical processes, and its essence hasn't changed for thousands of years. The ingredients for a basic soap are simple. Just cook up some fat with a base—usually sodium hydroxide, commonly known as lye. Chemically speaking, fats are triglycerides, meaning that three fatty acids are attached to a backbone of glycerol. Hydroxide ions attack the molecules of fat, breaking them down into fatty acids and glycerol. The sodium salt of the fatty acids that form is what we call soap. One end of this molecule dissolves in fatty or oily substances, the other in water. When we rinse, the oils and whatever dirt is embedded in them are washed away. If the fat used is beef tallow, as it most commonly is, the term for that particular soap is sodium tallowate. Lard or vegetable oils can also be used. Palmolive soap, for example, is so called because it is made from palm oil.

The glycerol that is formed as soap is made can either be left in the soap, where it acts as a skin softener, or removed and sold separately for use in other cosmetic products as a moisturizer. Since lye is very irritating to the skin, it is critical that none is left in the soap. This is accomplished through superfatting, which means that the amount of lye added is not enough to react with all of the fat. The leftover fats also make the soap easier to cut and make it feel smoother on the skin. Reaction of lye with the fats in soap actually goes on for several weeks, even after the bars are wrapped. By the time soap reaches consumers, there is no risk of irritation from lye. Perfumes and dyes added during saponification, however, can cause skin sensitivities in susceptible people.

Titanium dioxide is often added to make soaps opaque, and tetrasodium EDTA is added as a chelating agent, meaning that this chemical has the ability to bind metal ions. Calcium or magnesium commonly present in water react with soap to form a scum, but this can be reduced by using EDTA. Soap is made industrially in large iron kettles, and traces of iron in the soap can lead to discoloration. EDTA takes care of this as well by tying up the iron. Dyes and perfumes in soap can react with oxygen, especially in the presence of metal ions. This problem is also reduced with the use of EDTA.

What common consumer product has difluoroethane as its active ingredient?

Gas dusters used to clean computers and electronic equipment. A blast removes dust, but it can also give the person using the product a blast if they deliberately inhale it. The euphoric effect can extract a stiff price: death.

Although such products are often referred to as "canned air," they actually contain compressed difluoroethane. At room temperature and pressure, difluoroethane is a gas, but it can be compressed to a liquid more easily than air or hydrocarbons such as butane, which have also been used in such dusters. Essentially, this means more blasts per can.

Although difluoroethane is flammable, it is much harder to ignite than butane, making it a safer product in this regard. And unlike chlorofluorocarbons—or CFCs, as they are known—which at one time were commonly used propellants in cans, difluoroethane has no ozone-depleting potential. But it does have the potential to cause harm when inhaled. Since difluoroethane is heavier than air, it displaces air from the lungs, meaning that brain cells are deprived of oxygen, which partially explains the mind-bending effect.

There is more, though. Difluoroethane is fat soluble and dissolves in cell membranes, affecting the way that cells communicate with each other. It also may have an effect on the activity of a specific stimulatory neurotransmitter called gamma-aminobutanoic acid, or GABA. Similar effects have been attributed to sniffing butane, or toluene, the solvent in many glues. Inhalation anaesthetics such as enflurane or halothane also function in this fashion. All these chemicals, including difluoroethane, can cause an irregular heartbeat, which can be lethal. And this isn't just theory; numerous deaths have been attributed to sniffing inhalants. In one well-publicized case, a fifteen-year-old boy was found dead in his bedroom, still clutching a can of duster. In another, a twenty-four-year-old woman lost control of her car and died just after having inhaled some gas duster. To prevent such incidents, the manufacturers have taken to adding a chemical with an extremely bitter taste to their products in the hope that this will prevent any experimentation with inhaling the difluoroethane.

Why is carboxymethylcellulose added to laundry detergent formulations?

To prevent dirt from redepositing on the items being laundered. Detergents are very effective at removing dirt from fabrics, but once the dirt has been removed, much of it gets dispersed in the wash water as very fine particulate matter. Water is not very effective at keeping particles suspended, meaning that the dirt can readily redeposit on clothes. Redeposition of soils is the major cause of fabrics greying over repeated launderings. But there is a solution to this problem. A small amount of carboxymethylcellulose, or CMC, keeps the dirt in suspension and prevents it from being redeposited.

Carboxymethylcellulose was first made by German chemists in 1936 to serve as an adhesive for paper products. It is soluble in water and can readily glue surfaces together as the water evaporates. After World War II, American scientists took stock of all the chemicals that were being produced by the German chemical industry and found that Henkel had been manufacturing large amounts of carboxymethylcellulose. They weren't quite sure what uses this material was being put to, so they began their own investigations. They quickly discovered that it was a marvellous suspending agent. When a bag of dirt was dumped into a cylinder containing about 1 per cent carboxymethylcellulose, it stayed suspended almost indefinitely, while in regular water the dirt sank to the bottom almost instantly.

Before long, carboxymethylcellulose began to be incorporated into laundry detergents. The magic behind CMC is its ability to modify the viscosity or thickness of a liquid. Just a slight increase in water viscosity is enough to prevent dirt from settling. A higher concentration of carboxymethylcellulose can create a gel. Toothpastes, K-Y Jelly, artificial tears, cosmetic creams and water-based paints make use of this thickening property. Since carboxymethylcellulose is completely non-toxic, it can even be used in the food industry to thicken ice cream or improve the texture of dietetic products.

Obviously, CMC is a useful substance, but what is its environmental footprint? Production requires the use of chloroacetic acid, which in turn requires chlorine. There is always some degree of concern about the use of chlorinated substances because of the possibility that they may wind up in incinerators, where they can give rise to dioxins. This, however, is not an issue in carboxymethylcellulose production, because the chlorine used ends up in the form of sodium chloride, or salt. This can be recycled to produce more chlorine.

Ketchup can be made to flow when it is shaken, but solidifies when allowed to stand. What scientific term describes this property?

Thixotropy. Who hasn't been frustrated by the unwillingness of ketchup to emerge from its bottle? But the solution is simple: a good shake, and it pours like a liquid—a *thixotropic liquid*. Just what happens isn't completely understood, but the violence somehow eases the molecular entanglements, allowing the gel to liquefy, at least until the motion ceases.

Ketchup isn't the only substance that exhibits such behaviour. Quicksand is another example. Thrashing about in it increases the danger of sinking because the movement causes the sand-water mixture to become more liquid. But perhaps the most curious example of thixotropy is the twice-yearly liquefaction of the dried blood of St. Gennaro, patron saint of Naples.

Gennaro was reputedly a third-century bishop who was beheaded by the Romans, although there is doubt about his very existence. In any case, a sample of his supposed dried blood appeared in 1389, along with stories about it miraculously liquefying when taken on a religious procession through the streets of Naples. But the liquefaction did not always occur, and it was purported that when this was the case, some sort of disaster would follow.

So, what's going on here? A spectroscopic analysis of the sample, which involved exposing the closed vial to light and determining what wavelengths were absorbed, has indeed shown that it probably is blood. But what causes the unusual behaviour? Professor Luigi Garlaschelli of the University of Pavia thinks he may have found the answer. The clue was that during the religious procession, the act of checking whether liquefaction has occurred involves repeatedly inverting the glass-walled portable relic case. This suggested that the substance inside may be thixotropic, but that is not a property associated with congealed blood. However, what if some substance had been added to the blood?

Garlaschelli and colleagues found that combining solutions of iron chloride and calcium carbonate with salt in a specific fashion yielded a dark brown thixotropic mixture that very closely resembled the "blood" of St. Gennaro. It turns out that ferric chloride can be found in abundance on active volcanoes such as Vesuvius, which is near Naples. Whether the blood of St. Gennaro has been doctored in this way will never be known, because the Catholic Church is opposed to opening the vial. But let's ask this question: What is more likely—that there is a God who gives the faithful information about their future through the miracle of the liquefaction of the blood of a saint who may or may not have existed or that some clever alchemist concocted a thixotropic mixture from some readily available substances?

The opposite of thixotropy is isotropy. This is when a fluid becomes more firm when agitated. A mixture of corn starch and water can be walked upon—as long as you walk quickly.

What common household item contains calcium silicate as an additive, and why is it there?

Salt, which is hygroscopic, meaning that it easily picks up water. This causes the crystals to "cake" together, preventing free flow. In 1911, the Morton Salt Company found an answer to this problem by adding magnesium carbonate, which has the capacity to absorb many times its weight in water without dissolving. This keeps the salt dry and allows it to pour freely. The company promoted this innovation with the slogan, "When It Rains, It Pours." Today, the anti-caking agent used is calcium silicate, because it is even more impressive in terms of the amount of water it can absorb—about six hundred times its weight. Obviously, not much calcium silicate has to be added; in fact, less than one-half of 1 per cent by weight is needed. However, calcium silicate is not water soluble, so it may cloud that water and settle at the bottom of some canning recipes (which is an aesthetic issue but not a health problem).

What is the most commonly used plastic?

Polyethylene. You carry your groceries and dry cleaning home in it, you store your leftovers in it, you squeeze your shampoo out of it. Polyethylene is made by joining together molecules of ethylene, which in turn are obtained from petroleum. Just how these fundamental units are linked together determines the final properties of the product. If the ethylene units are joined in long chains with many branches stemming from the main chain, we have what is referred to as low-density polyethylene. This was the first type of polyethylene manufactured in the 1930s, and while it wasn't very strong, it was useful for plastic wraps, squeeze bottles and even plastic flowers.

About twenty years later, Karl Ziegler and Giulio Natta developed catalysts that allowed the ethylene units to be joined in long chains without any branching. The resulting high-density polyethylene was much stronger and had a higher melting point. It was ideal for toys, gasoline tanks, radio and television cabinets and myriad other items. It could even be fabricated into liners for surgical gloves, making them fifteen times as cut resistant.

Polyethylene is an excellent electrical insulator, which made it tremendously useful during World War II. It was used by the Allies to coat electrical cables in radar installations. Sir Robert Watt, the inventor of radar, paid it homage: "The availability of polyethyelene transformed the design, production, installation and maintenance of airborne radar from almost insoluble to the comfortably manageable. A whole range of aerial and feeder designs otherwise unattainable was made possible, a whole crop of intolerable air maintenance problems was removed. And so polyethylene played an indispensable part in the long series of victories in the air, on the sea, and on land, which were made possible by radar."

Alabaster is a mineral that can be carved into objects ranging from bottles to lamps. There are actually two kinds of alabaster: one composed of calcium sulphate, the other of calcium carbonate. What chemical test could you use to distinguish between the two?

Add a drop of a strong acid such as hydrochloric, sulphuric or nitric. If the alabaster is made of calcium carbonate, there will be a reaction. The mineral will effervesce, giving off telltale bubbles of carbon dioxide. This test is general for carbonates, which will always release carbon dioxide when reacting with an acid. Neutralizing stomach acid with Tums is a classic example. Hydrochloric acid in the stomach reacts with the calcium carbonate in the tablet, resulting in the evolution of carbon dioxide. *Burp!* In the process, the hydrochloric acid is neutralized. Alabaster made of calcium carbonate is also known as calcite alabaster and was probably first used by the ancient Egyptians, who fashioned bottles and even sarcophagi out of it. Alabaster composed of calcium sulphate is much softer and will not react with acids. It is also referred to as gypsum alabaster and is found extensively in England.

What is the oil most commonly used in oil paints?

Linseed oil, also sometimes called flax oil. The oil is extracted from the seeds of the flax plant, the same plant that yields the fibres used to make linen. Today, when people hear about flaxseeds, they think of their role in nutrition and health. Indeed, flaxseeds have laxative properties, can lower cholesterol and also contain lignans, which have anti-cancer effects. But up to the 1950s, when oil paints began to give way to modern latex paints, the oil from linseeds was the prime "drying oil" used in paints.

All paints have two basic ingredients: a "vehicle," which provides the film that covers a surface, and a pigment, which is composed of tiny particles suspended in the vehicle to produce colour. The vehicle in the early oil paints was linseed oil, which upon exposure to air produced the film that coated and protected the surface. Linoleic acid, a type of fatty acid we refer to as polyunsaturated because it contains multiple double bonds in its molecular structure, is the key ingredient in linseed oil. Its double bonds provide sites in the molecule that can react with oxygen, the net result being the formation of a three-dimensional network in which linoleic acid molecules have cross-linked to make a tough film.

Originally, oil paints were formulated with linseed oil, a thinner such as turpentine to allow easy application, and a pigment like lead carbonate. After the paint was applied to a surface, the turpentine evaporated and the linoleic acid reacted with oxygen to produce the required film. Such oil-based paints took a long time to dry, reeked of turpentine and were, in general, not suitable for metal surfaces. They have, by and large, been replaced by "plastic" or latex paints, which are formulated using an emulsion of synthetic polymers in water. In this case, the polymers that characterize the needed film are already present in the paint; no polymerization occurs after application. Polyvinyl acetate, polymethyl methacrylate and styrene-butadiene resins are suspended in water along with pigments. After application, the water evaporates and leaves a layer of paint behind.

The reaction that made oil-based paints so useful is exactly the type of reaction that restaurants want to avoid in their fryers. Reaction of unsaturated fats with oxygen reduces the length of time an oil can be used. Hydrogenating such fats gets rid of some of the troublesome double bonds, but unfortunately results in the formation of the notorious trans fats.

ANIMAL
PHARMACY

In 1910, irate medical students from University College in London attacked a statue in a park, precipitating a full scale battle with the police. What did the statue depict?

A brown dog. The statue had been commissioned by the World League Against Vivisection as a memorial to the animal at the centre of a celebrated British slander trial of 1903. A suit for libel was brought by University College physiologist Dr. William Bayliss against barrister Stephen Coleridge of the National Anti-Vivisection Society, who in a public speech had accused Bayliss of mistreating a brown terrier in a demonstration to medical students.

Performing experiments on live animals in the pursuit of medical knowledge was standard fare at the time, and Bayliss, along with colleague Ernest Starling, were at the forefront of physiological research. Using dogs, they had shown that the entry of food into the intestine stimulated the release of digestive juices from the pancreas even when intestinal nerves were severed. This clearly showed that the signal to the pancreas was passed through chemicals in the bloodstream and not through nerves. Starling coined the term "hormone" from the Greek word for "arouse," for such chemical messengers capable of stimulating distant organs even though secreted in very small quantities. He isolated and identified "secretin" as the hormone that stimulates pancreatic secretions.

Such research was opposed by anti-vivisectionists, who claimed that animals were being abused. The issue came to a head in Britain when two Swedish anti-vivisectionists enrolled as students at the London School of Medicine for Women. This school did not perform animal experiments, but its students had visiting rights to other medical schools. And it was during a lecture at University College that they witnessed what they claimed was an experiment by Dr. Bayliss on the salivary glands of a dog that had not been anaesthetized. Furthermore, Starling had performed experimental surgery on the pancreas of the same animal, and according to the laws at the time, animals were not supposed to be subjected to more than one experiment.

The two ladies brought the "affair of the brown dog" to the attention of Stephen Coleridge, who proceeded to give a series of inflammatory public speeches against Bayliss and triggered a lawsuit—probably on purpose to gain publicity for the anti-vivisection movement. There was indeed a great deal of publicity, but the jury found for Bayliss, who had provided evidence that the dog was actually anaesthetized and not maltreated. He was awarded two thousand pounds and three thousand in court costs, a staggering amount for the times.

In response to what they believed to be an unfair judgment, the World League Against Vivisection convinced the borough of Battersea to erect a bronze statute of the dog sitting on top of a granite pillar. Battersea was an appropriate location because it housed the Battersea General Hospital, which refused to perform animal experiments or hire doctors who engaged in the activity. The inflammatory plaque on the statue was regarded by medical students as slander on their chosen profession, culminating in the attempt to destroy the statue in 1910. Many were arrested, including one for "barking like a dog."

The "brown dog riot" resulted in round-the-clock police protection for the statue; this proved to be so expensive that one night in 1910 the statue was quietly removed and melted down. Antivivisectionist demonstrators demanded the return of the statue, but it was not to be—at least not until 1985, when a new brown dog statue was commissioned by the National Anti-Vivisection Society and now stands in a secluded corner of a park in Battersea. The plaque on it repeats the original accusations and adds, "In 1903, 19,084 animals suffered and died in British laboratories. During 1984, 3,497,355 animals were burned, blinded, irradiated, poisoned and subjected to countless other horrifyingly cruel experiments in Great Britain." Not a shred of evidence is provided for these allegations, and there is no mention of the countless human lives that have been saved by animal experimentation.

What animal did the ancient Greeks believe sweated blood?

The hippopotamus. The Greeks, it seems, were the first to note the thick red substance that commonly oozed from the skin of this large relative of the pig. It had all the appearances of blood, but at that time there was no way to investigate the chemistry of the substance. Not that getting close to a hippo in the wild to wipe sweat off its back was an attractive proposition. And so the chemistry of red hippo sweat remained a mystery until 2004, when Professor Kimiko Hashimoto at Kyoto University in Japan decided that the world could no longer wait for an answer.

A zoo was approached to see if they had a well-fed candidate that was unlikely to look upon a graduate student charged with wiping its back as a threat. Hippos are herbivorous but nevertheless dangerous to humans. Hippos are short-sighted and they instinctively try to clamp their huge jaws around anything they fear may be a predator. Over short distances they can reach speeds up to fifty kilometres an hour, easily outrunning a human. They are extremely territorial and protective of their young. They reputedly kill more people every year than lions or crocodiles.

It seems the Japanese zoo had a friendly creature that didn't mind having its back swabbed with gauze a couple of times a week for a few months. After purifying the sweat, Hashimoto and colleagues managed to isolate two compounds, one red and one orange, christening them hipposudoric acid and norhipposudoric acid. But why did the hippo produce these compounds?

A clue came from published accounts noting that albino hippos are healthy in spite of their rather thin epidermis being exposed to the blazing sun. Indeed, it turned out that the two coloured compounds absorbed some components of visible light, but of ultraviolet as well. Basically, hippos produce their own sunscreen! Although they spend much of their life in the water, they have to venture out to graze on the vast amount of grass they need to satisfy their hunger. But even in the water, their backs are exposed to the sun, so they need protection, which the coloured pigments deliver. And that's not all. Hipposudoric acid turns out to have antibacterial properties as well. Hippos in the wild are belligerent and are constantly nipping at each other. As a result, they are covered with a variety of wounds, and it seems evolution has endowed the animals with a natural antibiotic that keeps them from getting infections. There is yet another mystery concerning hippopotami: they are retromingent. In other words, they urinate backwards. Probably to mark territory.

How would you protect your dog from methylxanthines?

Hide the chocolates. Methylxanthines comprise a set of compounds found in chocolate, the best known of which are theobromine and caffeine. Although these are potentially toxic, consuming them is not a problem for humans because we have efficient ways of breaking down and excreting methylxanthines. But many animals, including dogs, cats, foxes, parrots and coyotes, are not equipped with our rapid detoxication mechanism and can be readily poisoned by chocolate. A small dog can be killed by as little as one hundred grams of unsweetened dark chocolate.

The toxicity of chocolate for animals can, however, come in handy when it comes to controlling animal populations. Coyotes, for example, are a huge problem in the U.S., killing millions of dollars worth of livestock every year. Fences don't stop them, and the traditional bait poison, sodium cyanide, is very dangerous to humans as well. Researchers at the U.S. Department of Agriculture may have found a solution to the coyote problem in chocolate. A mixture of theobromine and caffeine dispenses with coyotes very efficiently. Furthermore, the researchers found that the mixture works against slugs as well, potentially leading to a new class of safer pesticides. Indeed, for humans the safety profile of these chemicals is impressive. We would have to eat kilograms and kilograms of chocolate at a single sitting to experience toxicity. No need to try it, though.

A small German pharmaceutical company's logo resembles a flying bat. Why?

It produces an anti-vampire drug. Just kidding. It's testing a protein isolated from the saliva of the South American vampire bat in the treatment of strokes caused by blood clots.

Vampire bats feed on animal blood, and to avail themselves of a plentiful supply of nourishment they secrete a protein that prevents blood from clotting. They inject this protein into the wound when they bite into their victim in order to keep the blood flowing. Currently, tissue plasminogen activator is used to dissolve blood clots in stroke victims, but the hope is that the protein isolated from the saliva of *Desmodus rotundus* will be more effective.

Paion, the German company interested in marketing the substance, has isolated the gene responsible for the formation of the clot-dissolving protein and has been able to come up with a genetically engineered version called Desmoteplase, which is currently being tested in hospitals in Europe and the U.S. Perhaps Draculase would have been a better name.

The Florida cottonmouth is a poisonous snake. Researchers have hopes of using its venom in what sort of product that may one day be available in supermarkets?

Laundry detergent. Snake venoms often contain enzymes that prevent blood from clotting, making it easier for the venom to spread throughout a prey's body. There has been interest in using such enzymes to break up clots in the human body, especially since many heart attacks are triggered by blood clot formation. But bio-chemist Devin Iimoro at Whittier College in California had another idea. Perhaps the same enzymes that break down blood clots could be used to remove blood stains from fabrics. After all, the chemistry is very similar.

The coloured components of blood are bound to fabric by proteins called fibrins. If these can be degraded, the stain can be rinsed away. In preliminary experiments, fibrinolytic enzymes isolated from commercially available venom successfully removed most of a blood stain from white denim. Pretty interesting, but actually not an earthshaking discovery.

Papain, an enzyme extracted from pineapples and available as meat tenderizer, can do the same, as can a variety of products on the market that contain enzymes isolated from bacteria. None of these enzyme products, however, can be used on wool or silk, which are composed of proteins and would therefore also be degraded. In any case, 3 per cent hydrogen peroxide with a couple of drops of ammonia does a very good job of removing blood stains. Florida cottonmouths need not apply for positions in laundromats.

What is unique about the pitohui bird of New Guinea?

It is the only bird known to contain a toxin. And not just any old toxin, but batrachotoxin—which previously has been found only in the skin of some South American "poison dart frogs." That's the stuff that was sometimes applied to the tips of arrows to render them more deadly. The discovery of the poison in the bird was made back in 1989, when ornithologist Jack Dumbacher caught one of these birds in a net in New Guinea. When he inadvertently touched his lips after handling the bird, Jack noted that his tongue and lips went numb. This spurred a chemical investigation that revealed the presence of batrachotoxin in the bird's feathers.

But how did it get there? Recent investigations suggest that the birds, or indeed the frogs, do not synthesize the poison. They get it from their food supply—specifically, from a species of beetle. Melyrid beetles seem to be the source of the toxin. The theory will be tested by feeding toxic beetles to captive pitohui songbirds, carrying out a chemical analysis of their feathers and comparing the contents with the feathers of birds fed an alternative diet.

There are several species of pitohui (pronounced "pit-oo-eey"), with the most poisonous one being the hooded pitohui. Feeding just a few milligrams of its skin to a mouse will kill it in a few minutes. You can recognize the hooded pitohui by its coloration: its plumage is a brilliant red and black. The less poisonous birds are more brown coloured. Maybe the intense colour serves as a warning to predators to leave the bird alone.

Anyway, the moral of the story is that if you are wandering through the forests of New Guinea and you come upon a beautiful red and black pitohui bird, just enjoy its vocal talents from a distance. Do not attempt to handle it. This bird is definitely worth leaving in the bush.

WHAT'S
YOUR
TOXIN?

Poisoning by what substance leads to a condition called saturnism?

Lead. The term "saturnine personality" was introduced by astrologers in the fifteenth century to describe people born under the planet Saturn. These people were said to be of a gloomy or surly disposition. A saturnine humour was associated with frequent headaches, fatigue, irritability and depression. Little wonder that lead poisoning, which is characterized by these very symptoms, came to be called saturnism.

Lead has a long history of poisoning people. Citizens of ancient Rome suffered the consequences of lead exposure, although they didn't realize it. They drank water that flowed through lead pipes and drank wine from lead containers. Indeed, they favoured such wine because it was especially sweet. What the Romans didn't know was that acetic acid in wine reacts with lead to form lead acetate, a very sweet substance. Eventually, the Romans produced lead acetate on purpose, marking the birth of the artificial sweetener industry. Not an auspicious beginning, since lead acetate, which later came to be called sugar of lead, was highly toxic.

Lead was widely used by the ancients because it was easily extracted from its ores and was readily fashioned into items of great utility. Roman aqueducts were engineering marvels made possible by the use of lead pipes, and the easy access to water they provided greatly improved the quality of daily life. The introduction of lead pipes is an excellent example of the ambivalent nature of many scientific discoveries. There's a problem to be solved, such as acquiring fresh water, and a solution is found. But the solution often introduces a novel problem, which hopefully is a smaller one than that which was solved. However, it may take some time before the new problem is recognized. Efforts are then made to solve this problem, but *that* solution may introduce its own problems. When the toxicity of lead became apparent, lead pipes were eventually replaced with iron and plastic ones, with each of these introducing some issue such as corrosion or the leaching of plastic components into the water.

Most uses of lead—such as the addition of tetraethyl lead to gasoline to improve combustion, or its inclusion in paints to impart vibrant colours and moisture resistance—have been eliminated, but lead is still essential for the production of car batteries and the yellow markings that prevent accidents on our highways. And since lead is ubiquitous in nature, it will show up in trace amounts in numerous consumer items ranging from cosmetics to dietary supplements.

Much effort is being expended to reduce exposure to lead because of the metal's ability to attack the nervous system, weaken joints, decrease fertility, cause gout, decrease IQ and affect behaviour. Surveys have shown that anyone committing a crime of violence is likely to have something like four times more lead and aluminum in their hair then a normal person.

Human behaviour can be affected by lead, no doubt about it. Many artists in history may have suffered from the effects of lead paints. Francisco Goya appears to be one of these. His bouts of illness were often characterized by symptoms typical of lead poisoning. Curiously, one of his most famous paintings is *Saturn Devouring His Son*, which depicts the Roman god eating his son because of a prophecy that the son might one day usurp his throne. Indeed, lead can be described as a "devouring element."

What would you be accused of if you were told that you are "first cousin to an upas tree?"

Murder. This expression was used by Lord Peter Wimsey, Dorothy Sayers's famous fictional detective in *Unnatural Death*, written in 1927. The upas tree has a reputation for being highly poisonous; it's based mostly in fiction but has a scientific root.

The tree, known botanically as *Antiaris toxicaria*, is an evergreen native to Southeast Asia. Europeans first heard about the tree from an account by Dutch surgeon Nicolas Foersch, who encountered it in Java and published an intriguing narrative about its poisonous nature in 1873. It was certainly a scary account. "Not a tree, nor blade of grass is to be found in its vicinity, not a beast or bird, reptile or living thing." The inference was that the tree gives off a poisonous vapour that destroys all life.

Naturalist Erasmus Darwin, Charles's grandfather, took Foersch's fearsome description of the upas tree as fact and popularized its poisonous nature in a widely read book. Famed artist Francis Danby even produced a painting depicting a putrid stream of vapour rising from the tree. And a story made the rounds about how the tree's poison was collected by criminals who were given the task as a means of escaping execution. They had to wait until the wind was blowing away from them—and toward the tree—and then scamper to the tree and collect the poison from its sap before the wind changed direction.

All of this, of course, is fable, but the sap really does contain antiarin, a heart stimulant that is potentially toxic. And it was really used by natives of Java to poison the wells of Dutch colonists.

There is also an account, possibly true, of a local ruler in Java who had his unfaithful concubines executed by introducing the sap of the tree into an incision made on their bare breasts. It seems not to be a good idea to cheat on a powerful husband in Java. But you can safely stand next to an upas tree and admire it without risking your life.

Who would be most worried about the consumption of locoweeds?

Ranchers who raise livestock. Locoweeds are poisonous plants that appear to drive cows, horses and sheep crazy when they graze on them. The animals are said to have gone "loco," from the Spanish word for crazy. So, how can you tell if a cow has gone crazy, given that this creature has never ranked high in the intellectual hierarchy of beasts?

With a diet sufficiently high in locoweeds, cows will develop an irregular gait, dull and staring eyes, loss of appetite and a decreased libido. No surprise that staggering cows, staring into the void and disinterested in sexual activity are readily labelled crazy by humans. The affected animals also become lethargic and depressed, although it isn't clear how one makes that diagnosis in a cow, an animal not noted for particular euphoric activities in any case.

The culprit in locoweed is swainsonine, a compound actually produced by a fungus living inside the plant's cells. A variety of plants can harbour the fungus, so that locoweeds actually encompass a number of species, although the ones that are of most concern in North America mostly belong to the Astragalus or Oxytropis families. Signs of poisoning, such as emaciation and neurological damage, can appear after just a couple of weeks of grazing on the toxic plants.

Herbicides such as clopyralid, picloram and metsulphuron have been used by ranchers where locoweeds are widespread in forage fields, mostly in the western regions of North America. The locoweed problem is not insignificant; estimates are that toxicity in free-range livestock costs the industry about $100 million a year. There are no treatments for locoweed poisoning, but most animals recover when the toxic plants are removed from the diet.

Just how swainsonine carries out its nefarious activities isn't clear, but the evidence points toward interference with some enzymes involved in the metabolism of mannosides, commonly occurring carbohydrates in plants. As a result, cellular carbohydrate debris builds up in cells, causing what one researcher has inventively called "cellular constipation." Is there anything good that can come out of the study of locoweeds? Maybe. Some preliminary evidence indicates that swainsonine may have anti-tumour activity and that it may also reduce the toxicity of some anti-cancer drugs. Its appetite-suppressant activity is also worthy of investigation. While swainsonine may drive livestock crazy, it may eventually prove to be of some pharmacological value.

What good came out of the development of Lewisite, a blistering agent developed for chemical warfare during the First World War?

A drug called British anti-Lewisite, or dimercaprol, which turned out to be an antidote for poisoning by arsenic, mercury, lead and other heavy metals. Wilson's disease, which is an overload of copper in the body, is also treated with intramuscular injection of dimercaprol.

Lewisite was developed by American chemist W. Lee Lewis, based upon a reaction he had found in a thesis by Father Julius Nieuwland, who eventually became a professor of chemistry at Notre Dame University. Nieuwland had been interested in acetylene and eventually found a way to use it as a raw material to make synthetic rubber. During the course of his research, he had made an arsenic derivative that he noted was capable of causing severe blisters. Lewis exploited this discovery to make Lewisite, but production was too late for use during the war. Experiments with this "dew of death," as the liquid came to be called, proceeded during the 1920s, and some attempts to use it were made during World War II. It turned out be ineffective because the substance was readily broken down by humidity in the air. Also, its characteristic odour, which resembled that of geraniums, was a telltale sign alerting enemy troops to don gas masks.

Between the two world wars, Sir Rudolph Peters at Oxford University investigated antidotes for chemical warfare agents. Because of concerns that the Germans might also have developed agents such as Lewisite, he focused on how this substance worked. The toxicity turned out to be due to the presence of an arsenic atom in the compound that was capable of inactivating essential enzymes by binding to sulphur atoms in their structure. This gave Peters an idea: if he could come up with a substance that contained sulphur atoms the same way that enzymes did, maybe the Lewisite would bind to it and leave the enzymes unperturbed. Peters chemically modified a glycerol molecule by substituting sulphur atoms for the oxygens it contained and came up with "British anti-Lewisite." It never had to be used against Lewisite, but it turned out to have an important medical application. Not only was British anti-Lewisite capable of tying up arsenic, it also readily bound mercury, lead and other heavy metals. It is still stocked in hospitals to treat such poisonings.

In the eighteenth century, consumers could purchase wafers containing arsenic. What for?

They turned complexions white. Arsenic can destroy red blood cells, which transport oxygen. With fewer red cells, less oxygen is transported around the body and the skin becomes paler in colour. Paleness was a status symbol because it distinguished the higher classes from the poor peasants who tanned as they toiled out in the fields. Some unfortunate women died from arsenic poisoning in an attempt to be fashionably pale. At the same time, Signora Toffana's arsenic-based face cream was used in Italy. It is estimated that six hundred husbands died from too close a contact with Signora Toffana's product.

Why did Dr. David Jones, a chemist at the University of Newcastle, set out in 1980 to track down the colour of Napoleon's wallpaper?

He was interested in determining whether it contained arsenic. The cause of Napoleon's death on the island of St. Helena has been much debated, but a post-mortem did reveal the presence of arsenic in his hair. This suggested that he might have been poisoned by the British, who were holding him prisoner. Jones, however, thought there might be another explanation for the presence of arsenic.

In the nineteenth century, a pigment known as Scheele's green was commonly used in fabrics and in wallpaper. The bright green colourant, invented by the Swedish chemist Scheele, was actually composed of copper arsenite. Although arsenic was known to be highly toxic, there was no great concern over the use of Scheele's green because people were not in the habit of eating their wallpaper. But it seems they didn't have to eat it to be poisoned by it.

If wallpaper becomes damp, it can foster the growth of mould, and some moulds digest copper arsenite and convert it to a mixture of arsine, dimethyl arsine and trimethyl arsine, all of which are highly toxic gases that can be inhaled. A sample of the wallpaper eventually was found in a scrapbook of someone who had visited St. Helena in 1823. Testing revealed the presence of arsenic in the green pattern. This, of course, does not prove that Napoleon died of arsenic poisoning, but it does show that it was a possibility. Most medical experts, though, now believe the emperor died from stomach cancer.

In a buffet, you will often see dishes being warmed from underneath with a flame from a little can. What is the fuel that burns in such cans?

Alcohol—more specifically, a mixture of ethanol and methanol. The alcohol has been "gellified" by adding a thickening agent so that there is no spillage, and is sold in cans, most commonly under the brand name Sterno. Such cans can provide heat for several hours, but there have been some unfortunate incidents with the product. In a few instances, people desperate for alcohol have squeezed the liquid out of the gel and drunk it, not realizing that it contained toxic methanol. In 1963, thirty-one people in Philadelphia died in this fashion.

Why did California issue a health advisory about eating fried grasshoppers covered with chili—a traditional Mexican treat?

Worry about contamination with lead. Chili powder is at the heart of Mexican cuisine, and it finds its way into numerous foods ranging from candies and salsa to fried grasshoppers. Unfortunately, it seems that lead sometimes finds its way into the chili powder, occasionally in amounts high enough to cause toxicity, especially in children. This was discovered when routine testing of foods in California revealed high levels of lead in candies sold in Hispanic neighbourhoods.

These spicy candies, imported from Mexico, were made with guajillo chilis, but the source of lead was a mystery. An investigation was deemed to be in order because in some cases the lead content was well over the permissible level of 0.2 parts per million. Lead is an insidious poison; even small amounts are capable of causing kidney damage, behavioral problems and learning disabilities. Children are especially sensitive because lead accumulates in their still-developing skeletal and nervous systems. The problem of lead toxicity eased with the banning of lead-based household paints in 1978 and leaded gasoline in 1986, but other environmental sources of lead remain, such as the unusual case of chili peppers.

Where was the lead coming from? Investigators found two sources. When chilis are dried, they are spread out on concrete slabs, where they are exposed to dust and dirt. Many soils contain naturally occurring lead compounds, which may become a problem if they are concentrated. That is just what happened in the case of the peppers, which were never cleaned before being ground up into chili powder. But a secondary source was even more disturbing. Since farmers sell their chili peppers in burlap sacks to millers by weight, they sometimes add scrap metal, including lead, to increase the weight. The millers use magnets to remove the scraps, but lead is not attracted to magnets.

Chilis destined for foreign sales are cleaned, but apparently Mexican producers take many shortcuts where the domestic market is concerned, and it was candies produced for local consumption that found their way across the border into California. Hence the warning about candies and fried grasshoppers. Salsa did not pose a hazard, because in the manufacturing process the chilis are suf-ficiently diluted.

Our word "toxic" derives from the ancient Greek word *toxikon*. What does it mean?

Arrow. The Greeks were familiar with poisoned arrows, as is clear from the mythological story of Achilles. In one version of the tale, Achilles' mother, the immortal sea nymph Thetis, tried to make her son immortal by dipping him into the river Styx, holding him by the heel. Since the heel remained above the water, it was Achilles' only vulnerable spot, and it was that very spot that a poisoned arrow found in battle.

The Greek word for poison was *pharmakon*, and *toxikon pharmakon* was therefore the term used to describe a poisoned arrow. The Romans derived the Latin word for poison, *toxicum*, from this expression but basically took the wrong half of the phrase. Our word "pharmacy," of course, derives quite reasonably from *pharmakon*, since drugs improperly used can be poisons and poisons properly used can be drugs. The classic example would be curare, an arrow poison extracted from a South American plant. It is a muscle relaxant that can kill by paralyzing the respiratory system but can also be used in abdominal surgery to prevent muscle spasms.

Just what poison the Greeks used on arrows is uncertain, but they were certainly aware of plant poisons, as Socrates could testify with his hemlock experience. Taxines found in the leaves, bark and berries of the yew tree are known to interfere with heart function and are strong candidates for arrow poison. Homer described archers besieging the walls of Troy with yew bows, so the Greeks were undoubtedly familiar with such uses for this tree. It is quite possible that they had mastered the art of extracting poisons from its leaves. There is evidence of the knowledge of yew poisons in antiquity; Pliny, the Roman encyclopedist, wrote of poison ingested from kitchen utensils made of yew wood.

Underlining the relationship between poison and potion, taxol, derived from the Western yew tree, is used today in the treatment of some cancers. Without a doubt, the most effective arrow poison is the one used by the Choco people of western Colombia who dip their blowgun darts into the secretions of frogs of the genus Phyllobates. Batrachotoxin in these secretions is probably the most potent natural poison known. Its name derives from the Greek *batrachos*, for frog, and the chemical is so potent that one hundred micrograms, roughly the weight of two crystals of salt, can kill a human.

What potentially toxic substance has been linked with artificial turf?

Lead. The current versions of artificial turf come close to a grass-like feel. They even look like grass. The blades are made of polyethylene, woven into a polypropylene backing. The space between the ribbons of polyethylene is filled with fine rubber particles made by grinding frozen tires. Tiny amounts of lead have been detected as contaminants in the pigments used to dye the grass as well as in the rubber particles. In theory, dust from the rubber and the polyethylene can be inhaled or ingested. Although the levels of lead are very low, it is an insidious toxin that may cause neurological effects in trace amounts. It is unlikely that athletes playing on artificial turf are at risk of lead toxicity, but infants playing on such turf could conceivably put rubber particles in their mouth when parents set them down while watching athletic events. But artificial turf also has pluses when it comes to toxicity. No pesticides, herbicides or lawn mowers are required. The benefits outweigh the risks.

What happened in Seveso, Italy, in 1976?

Mechanical and human errors led to an industrial accident that resulted in the highest known exposure to the most toxic form of dioxin in a residential population. On July 10, 1976, a cloud of gas and dust escaped from a malfunctioning reactor in a small chemical manufacturing plant just north of Milan in the Lombardy region of Italy. The dust contained an estimated three kilograms of TCDD, or 2,3,7,8-tetrachlorodibenzo-p-dioxin, a notoriously toxic compound.

Seveso, a small town with a population of seventeen thousand in 1976, was the most directly affected community, so it became known as the Seveso Disaster. Exactly how much of a disaster it was is debatable. Chemical plant officials quickly informed the local authorities of the accident, but it took a week before the public was alerted. During this time, pets and animals began to die, and the hands and faces of the children who played with what they thought was snow began to break out with various lesions—a classic symptom of dioxin exposure.

The term "dioxin" actually refers to a group of compounds that share a common basic molecular structure, incorporating a varying number of chlorine atoms. TCDD, the most toxic version, features four chlorine atoms on the periphery of the molecule. It rose to fame as a contaminant in Agent Orange, the herbicide used during the Vietnam War to defoliate trees and reveal Viet Cong hiding places. This compound is never manufactured on purpose but is a contaminant formed during the manufacture of some industrial chemicals and can also be a natural byproduct of combustion processes that involve chlorine-containing organic material.

Besides causing a persistent form of acne called chloracne, TCDD can also affect the functioning of nerve cells, can lead to thyroid disorders, and can damage the immune system. In animals, TCDD has been shown to cause birth defects and cancer. Whether these effects can also be expected in humans is uncertain due to a lack of controlled studies. Nevertheless, of the 150 pregnant women who were exposed to TCDD in Seveso, thirty decided to have abortions. None of the fetuses showed signs of defects, and of all the babies that were carried to term, only two exhibited birth defects, well within the number normally expected.

Eventually, six hundred people were told to leave the Seveso area, and two thousand more had their blood tested. Despite the heavy exposure to dioxin, no one died, indicating that humans, unlike animals, may not be acutely sensitive to TCDD. Eighty thousand animals in the area around Seveso were slaughtered to prevent dioxins from entering the food chain. A year after the accident, an epidemiological investigation of some 220,000 people in the area began under the auspices of the International Steering Committee, staffed with renowned experts from all over the world. In February 1984 the committee released its final report, stating that "with the exception of chloracne, no ill effects can be attributed to TCDD."

Since then, researchers have continued to monitor the health of the exposed population and have found a small increase in some rare cancers, but when all cancers are pooled, no statistically significant excess has been observed. There has, however, been one documented consequence of the Seveso accident: a higher proportion of female babies than statistics would dictate have been born.

What does farmed puffer fish not have?

The potent poison tetrodotoxin. The puffer fish has long been a Japanese delicacy in spite of the fact that, if improperly prepared, it can be lethal. In the wild, puffers ingest bits of worms, shellfish and starfish that can harbour bacteria capable of producing tetrodotoxin. The toxin serves to protect the puffer from larger, predatory fish, presumably by decimating their population.

Tetrodotoxin is a remarkable poison—less than one milligram is lethal to an adult. To put this into context, there is enough poison in one puffer to kill thirty people. Only the batrachotoxin produced by the golden dart frog of South America is more potent than tetradotoxin. That one is so powerful that chickens and dogs have died after contact with paper towel on which one of the frogs had hopped.

The mechanism of tetrodotoxin's toxicity has been well established; it interferes with the way messages are conducted along nerves. Transmission of a signal depends on the passage of sodium and potassium ions in and out of cell membranes through what are known as ion channels. Tetrodotoxin blocks the sodium channel, and if present in a high enough dose, it stops nerve cells from functioning. Effects begin with deadening of tongue and lip sensations, followed by a prickly sensation that can spread all over the body, finally resulting in muscle paralysis. Death ensues when the muscles of the diaphgram become paralyzed, resulting in suffocation.

Since the transmission of pain signals involves the same process, a Canadian company, Wex Pharmaceuticals, has been exploring the possibility of using tiny doses of tetrodotoxin to treat severe pain. Millions of people who suffer from cancer do not get adequate relief from the currently available drugs, and Wex hopes that Tectin, the trade name for tetrodotoxin, will help some of them. So far, trials have not produced results as significant as the company had hoped.

While pharmaceutical companies are interested in fish that contain tetrodotoxin, puffer fish eaters look for fish that have had the toxin removed. Fugu, as the prepared fish is called, has had its liver, gonads, intestines and skin expertly removed by highly trained chefs. Still, trace amounts of tetrodotoxin remain in the flesh, leading to minor tingling sensations and a degree of excitement when eating fugu.

Over the years, hundreds of people have died from eating improperly prepared fugu, and many, like Homer Simpson, have come close to death but survived. The risk has given birth to the puffer fish farming industry in Japan. When the feed of the fish is rigorously controlled to eliminate the toxin-producing bacteria, fugu can be enjoyed without risk. Even the liver, which the Japanese claim is superior in flavour to foie gras, can be eaten.

Raising farmed puffer fish, though, is not without controversy. There is a rather large Japanese industry that supplies restaurants with wild-caught fugu after removing the fish's toxic organs and is obviously threatened by safe, farmed puffer. Spokespeople argue that the farmed variety cannot be guaranteed to be devoid of tetrodotoxin and that the farmed fish actually represent a risk. Evidence, however, argues against this position. On the other hand, true fugu devotees don't favour farmed fish because some of the appeal of eating fugu comes from the risk of death. They enjoy the tingling sensation as it rises up the arms and legs, and when it stops at the elbows and knees, they get ready to enjoy an outstanding meal. If the sensation goes beyond the elbows and the knees, they might as well still enjoy the meal, which will be their last one.

BURNING
ISSUES

In 1766, Henry Cavendish discovered a gas he called "inflammable air." What was this gas, and how did he make it?

The gas was hydrogen, which Cavendish made by adding pieces of zinc to hydrochloric acid. High school students still initiate this reaction to study the properties of hydrogen. Perhaps the most impressive property is hydrogen's flammability. Stick a glowing splint into a test tube of hydrogen and you get a nice pop. Of course, if you have a large amount of hydrogen, you get more than a pop—think of the explosions of the *Hindenberg* and the space shuttle *Challenger*.

Cavendish was a strange man. He looked like the proverbial mad scientist. His clothes were shabby—the crumpled violet suit, frilled cuffs and three-cornered hat spoke of a bygone era. His voice was shrill and he never looked anyone in the eye. Yet Henry Cavendish was one of the greatest scientists who ever lived. Born in 1731, he lost his mother when he was only two years old. He was raised by his chemist father, whose professional skills were worthy of praise from the great Benjamin Franklin. Young Cavendish spent four years at Cambridge University but never took his degree—he simply could not face his professors for an exam. For the rest of his life, he would have trouble communicating with people, but in the laboratory his talent was unparalleled.

Cavendish inherited a fortune and never had to worry about working for a living. In fact, at the time of his death he was the largest depositor in the Bank of England. So the young scientist could have lived like a king, but instead led the life of a recluse, using the inheritance to fund his scientific work. Henry's social life was apparently nonexistent. Although he had problems facing men, he would occasionally attend the scientific functions of the Royal Society. But women were a different story. Even the maids in his house were instructed to communicate with him only through notes and to stay clear of him. When one day Cavendish accidentally met a maid on the staircase, he had a back stairway built that he alone could use.

Cavendish worked tirelessly in his home laboratory, but it was his discovery of "inflammable air" that brought him lasting fame by changing the course of chemical history. He noted that when he burned hydrogen in a closed container, water was produced. Water, therefore, was not an element; it could be made in the laboratory. This was the final nail in the coffin of Aristotle's theory that everything was composed of air, earth, fire and water. Now the path was clear for the progress of chemistry.

What element derives its name from the ashes left behind when wood burns?

Potassium. Until the advent of gas and electricity, cooking was done in pots heated over a wood fire. The ashes left behind after the fire was extinguished were accordingly referred to as "pot ash," with the term eventually entering the English language as the single word "potash." Extraction of the ashes with water yielded a white crystalline substance that was eventually named potassium carbonate after it was found to contain the element potassium, first isolated by the great English chemist Humphry Davy back in 1800.

Potash has been used since antiquity in the manufacture of soap and glass, and was also essential for making gunpowder. Its most important use eventually turned out to be as fertilizer, since all plants need potassium for growth. Until the 1860s, potash was derived from the ashes of hardwood trees, mostly using a method patented by Samuel Hopkins of Vermont. In fact, the patent registered in 1790 is recognized as the first patent ever granted in the United States. Hopkins had discovered that heating the ashes for a second time in a furnace before dissolving them in water resulted in a higher yield of potassium carbonate. The crystals also were purer.

The reason was that carbon in the ashes, which was responsible for the dark discoloration of the potash, was converted to carbon dioxide by the heat, and this in turn increased the yield of the carbonate ion needed to form potassium carbonate. After about 1860, natural potash deposits were discovered in Germany, meaning that potash could be mined instead of being produced by burning wood. Today, potash is as important a commodity as ever, mostly for the production of fertilizer, and is supplied by potash mines located around the world.

Up to the 1950s, projection booths in movie theatres were required to be equipped with a toilet. Why?

Building codes required it so that the projectionist would not need to leave the booth. The concern was about fire. Movie film was made of cellulose nitrate, an extremely flammable material. Projectors used carbon arc lamps that got very hot and were known to set the film aflame. That's why a projector's supply and take-up reels were surrounded by fireproof magazines, and the projection and observation ports in the wall of the projection booth had steel fire shutters that closed automatically in the case of fire. In the 1950s, cellulose nitrate film was replaced by cellulose acetate, which is much less flammable and is referred to as safety film. Some old nitrate films are still around, but modern projection booths are not equipped for them. That's why signs are sometimes still seen on the projection booths of modern theatres reading, "Safety Film Only—This Booth Not Approved for Nitrate Film."

Fires were commonly used as signal beacons atop the Great Wall of China. What substance was commonly added to make the fire more intense?

Potassium nitrate, more commonly known as saltpetre. For combustion to occur, a source of oxygen is required. This is usually supplied by air or by a chemical that can liberate oxygen. Potassium nitrate is such a substance and is therefore referred to as an oxidizing agent.

Nobody really knows how the ancient Chinese discovered this property of saltpetre, but it could well have come about as a result of burning manure for fuel. When certain bacteria feed off organic waste they produce saltpetre, which becomes evident as a white deposit in manure piles. Adding these crystals to a fire increased the intensity of the flames.

Eventually, this observation led to more than just an improvement in the signalling fires on the Great Wall. Combining saltpetre with sulphur and charcoal resulted in gunpowder, which was one of the most significant of all Chinese contributions to society. That saltpetre is definitely of Chinese origin is indicated by the names given to this material by the Arabs, who called it "Chinese snow," and the Persians, who called it "salt from China." Europeans eventually learned that saltpetre could be produced from rotting organic matter, especially meat and urine. Saltpetre companies hired "petermen" who searched for abandoned dung heaps from which potassium nitrate could be extracted. The demand for this substance, however, grew so large that artificial nitre beds became a feature of life. Layers of decaying organic matter, old mortar and earth were piled in a compost heap and sprinkled regularly with blood and/or urine. In about two years, crystals of potassium nitrate appeared.

Recognizing the importance of saltpetre, King Charles I in 1626 ordered "his loving subjects to carefully and constantly keep and preserve in some convenient vessels or receptacles fit for the purpose, all the urine of man during the whole year, and all the stale of beasts which they can save and gather together whilst their beasts are in their stables and stalls, and that they be careful to use the best means of gathering together and preserving the urine and stale, without mixture of water or other thing put therein. Which our commandment and royal pleasure, being easy to observe, and so necessary for the public service of us and our people, that if any person do be remiss thereof we shall esteem all such persons contemptuous and ill affected both to our person and estate, and are resolved to proceed to the punishment of that offender with what severity we may." In other words, there would be penalties for mis-peeing.

How does throwing baking soda on a fire extinguish it?

When baking soda, or sodium bicarbonate, is exposed to a high temperature, it decomposes to form sodium carbonate, water and carbon dioxide gas. Carbon dioxide is heavier than air and displaces oxygen from the vicinity of the fire, thereby extinguishing it.

Three criteria have to be met for sustained combustion. A fuel and a source of oxygen are needed, and a high enough temperature has to be maintained for the fuel and oxygen to react. Gasoline in the presence of air will not ignite until a spark raises the temperature high enough for the reaction to begin. Once combustion starts, the reaction generates enough heat to sustain it. Water can be used to extinguish fires because it lowers the temperature to a point where combustion can no longer occur. Displacing oxygen also works— that's why throwing a blanket on a fire can extinguish it. Carbon dioxide also throws a blanket on a fire—a blanket of gas that does not support combustion. Many fire extinguishers contain compressed carbon dioxide, but sodium bicarbonate can also do the job. But there's a catch: carbon dioxide can't be used on all fires. In the case of combustible metals such as magnesium, sodium, titanium, zirconium or potassium, carbon dioxide is useless. It actually reacts with the metals, producing even more heat. Magnesium burns in a carbon dioxide atmosphere to yield magnesium oxide and carbon along with a great deal of heat. Water can't be used on such fires, either. With magnesium, for example, it reacts to yield magnesium hydroxide, hydrogen and heat. Then the hydrogen can ignite, making the fire even more intense. Sand can be used to extinguish such fires.

DRINKING
PROBLEMS

How do you prevent anthocyanins from colouring *blanc de noirs*?

By removing the grape skins after red grapes are pressed. Many people believe that red wine is made from red grapes and white wine from green grapes. They are wrong. When grapes are pressed, the juice is white, regardless of the colour of grape. But if the juice is left in contact with red grape skins, pigments leach out of the skins and colour the wine. These pigments belong to a family of compounds called anthocyanins, which are responsible for the coloration of many fruits and vegetables.

If red grapes are pressed and the juice is allowed to ferment after being separated from the skins, the result is white wine, referred to as *blanc de noirs*, meaning a white wine from dark grapes. A well-known example is a variety of Champagne produced from pinot noir grapes. Most white wines, though, are made from white grapes, with the skins being removed before fermentation starts. Since there is no need for colour, there is no point in having the juice remain in contact with the skins. Such contact would only serve to leach out tannins, which add astringency and a bitter flavour to the wine.

In the case of red wines, contact with the skins is necessary to infuse the fermenting juice with anthocyanins. But this means that tannins are also being leached out. So a prime question in red wine production is when to transfer the wine from the fermentation vessel, where it is in contact with the skins, to barrels, where it is not. Leaving the juice in contact with the skins after colour has fully developed is a no-no because it can yield too high a tannin content.

Now, tannins are not totally undesirable. They add body to the wine and a certain "pucker factor" that contributes to the wine-drinking experience. Tannins are complex polymers of molecules called phenols, and they can make saliva more viscous by linking together some of its naturally occurring proteins. Indeed, tannins extracted from plants are used to "tan" leather, a process that cross-links proteins, turning soft animal skin into a substance tough enough for shoes, belts and furniture.

Vintners know that colour is a marker of the fermentation process and have traditionally used its development to judge how long to let grape juice sit in contact with skins. Early experiments attempting to put such decisions on a scientific footing resulted in the production of "wine colour cards" that featured a series of circles in different red hues to be compared with the colour of a fermenting batch of wine. The colours were based on experiments that had shown the optimal time for transfer of the fermenting juice to barrels. Today, various sophisticated scientific instruments are available to help vintners produce high-quality wine of consistent flavour. Spectrometers can measure subtle differences in colour, while pressure transducer sensors can monitor the extent to which sugar has been converted to alcohol. Amazingly, though, the best instrument for distinguishing the subtleties of wine is still the human nose, which can detect the presence of some chemicals in concentrations that defy detection by the most sensitive instruments.

Why are some home brewers experimenting with adding alpha-galactosidase to their beer while it is fermenting?

To try to cut down on another type of fermentation—the sort that takes place in the colon and causes gas production. These inventive brewers are trying to avoid the dreaded scourge of "beer flatus" by doping their brew with the enzyme alpha-galactosidase, which has been commercially marketed as Beano.

Alpha-galactosidase is a digestive enzyme that helps break down complex carbohydrates into simple sugars. When not enough of this enzyme is present in the digestive tract, ingested complex carbohydrates pass through the small intestine all the way to the colon, where they become a tasty meal for the resident bacteria. The problem is that these are flatulent bacteria. When they digest the complex carbohydrates that have come their way, they produce a variety of gases that eventually trumpet their presence as they exit the body. These pesky complex carbohydrates are found in foods like cabbage, cauliflower, onions, oats, nuts and beans. But there is an answer.

Back in 1991, Alan Kligerman, who had studied dairy science at Cornell University, tackled the problem. He had already made a name and fortune for himself by marketing lactase, the enzyme needed to digest lactose, the main sugar in milk, and thereby endeared himself to the multitudes suffering from lactose intolerance. Now Kligerman came to the aid of the throngs afflicted by gassiness after eating foods harbouring the troublesome complex carbs, especially beans.

Kligerman hypothesized that alpha-galactosidase would help, and found in the fungus *Aspergillus niger* a readily available source. Indeed, trials soon showed that adding the enzyme, now christened

Beano, to the potentially offensive foods just before consumption greatly reduced gassiness by breaking the complex carbohydrates down into simple sugars readily absorbed from the small intestine.

Some of the complex carbs that cause the gas problem are also found in barley and are not broken down by yeast during fermentation. The result can be flatus-producing beer. Adding Beano to the brew can help with this problem. But it also affects the taste and alcohol content of the beer, since the complex carbs add body and flavour. This is altered when they are broken down to simple sugars. And the simple sugars can now be fermented by the yeast, so the alcohol content increases. Still, flatulent beer drinkers welcome this biochemical intervention.

Kligerman is not resting on his laurels, and has gone on to develop a product for gassy dogs. Whether their human masters will welcome this is unclear. After all, it would make it harder to blame the dog for human indiscretions.

Why was root beer–flavoured Kool-Aid removed from the market in the late 1950s?

It contained safrole, a compound the U.S. Food and Drug Administration banned because it was shown to cause cancer in rodents. Charles Hires was the first to concoct root beer back in the 1870s by blending a variety of root, herb and berry extracts that included juniper, wintergreen, sarsaparilla, hops, ginger, licorice, dog grass and birch bark. But eventually sassafras became the key flavouring. One of its main components was safrole, the compound the FDA declared to be a carcinogen in the 1950s, resulting in the ban of the compound from use in any food or beverage.

The FDA's ban of safrole caused a major trauma among root beer producers, who then had to scramble to find a flavour replacement. Root beer flavouring was not the only commercial use of sassafras. The bark of the tree was an important American export to Europe. Word had it that sassafras poultices were useful in treating sores and that extracts of the bark served as a "blood purifier." There is no evidence that safrole has any such useful therapeutic effect.

Needless to say, today's root beers are not flavoured with safrole and mostly use wintergreen as the dominant flavour. While the interest of root beer producers in safrole has faded, that of illicit drug manufacturers has increased. Safrole can be used as the starting material in the clandestine manufacture of methylenedioxymethamphetamine, better known as ecstasy. A few relatively simple chemical reactions can convert safrole into highly profitable ecstasy.

Of course, it is a challenge for crooks to get their hands on safrole since it is a controlled substance. Several chemical companies produce the compound, which is used in the commercial synthesis of such chemicals as the insecticide piperonyl butoxide. But any individual ordering safrole from a chemical company is likely to get a visit from the police. And they won't be after root beer.

Who first suggested that milk be pasteurized to make it safer for consumption?

No, it wasn't Louis Pasteur. Frans von Soxhlet, a German agricultural chemist, suggested in 1886 that milk sold to the public be heat treated. This process was eventually called pasteurization in honour of Pasteur's pioneering work on the destruction of microbes through heat treatment. Pasteur's area of interest, though, was wine and beer, not milk.

The fact that heat treatment made foods safer was known long before Pasteur, but the French chemist was the first to provide an explanation for the phenomenon. Pasteur realized that spoilage was due to chemical reactions initiated by living microbes, and that the reason heat treatment prevented spoilage was because of its destructive effect on these living organisms. If wine or beer turned sour, Pasteur maintained, it was because of contamination by acid-producing rogue yeasts after the alcohol-producing yeast had done its job. Heating of beer or wine would then destroy these invaders and preserve the beverage. Indeed, after the Franco-Prussian War of 1870, Pasteur, a noted French patriot, created his "beer of revenge," which would serve as a testimonial to the superiority of French brewing techniques over anything the Prussians could muster. Heat-treated French beer would keep indefinitely while Prussian beer would turn cloudy with time.

As far as beer went, this was mostly a cosmetic, not a health issue. But milk presented an altogether different scenario. Typhoid and scarlet fever, diphtheria, tuberculosis and various diarrheal diseases were all capable of being transmitted through the consumption of milk. It seems that until 1886, when von Soxhlet made his suggestion, nobody thought of pasteurizing milk on a large scale. And what a difference pasteurization made! In 1891, one in every four infants in New York City died, many from drinking tainted milk. This dropped to about one in fourteen when pasteurization was introduced. The advent of this life-saving technology, interestingly enough, was not without its critics. Some argued that heat treatment destroyed vital nutrients in milk and produced a "burnt" flavour. The controversy about the pros and cons of pasteurization continues to this day.

Milk contains some 100,000 naturally occurring compounds, and some of these certainly undergo chemical changes with heat. But that does not mean that these changes have any health consequence. There is a lot of bluster from raw-milk advocates about pasteurization impairing the nutritional merits of milk, but the arguments are not backed up by evidence. There is plenty of evidence, on the other hand, about unpasteurized milk causing disease.

While raw milk from an individual farm where cleanliness is scrupulously maintained is likely to be safe, milk pooled from many farms, as is the usual case today, may well harbour a range of bacteria. Until someone finds a way of preventing cows from defecating, pasteurization is the way to go. Heating milk to 72°C for fifteen seconds (so-called high-temperature/short-time treatment) or to 138°C for two seconds (ultra-high-temperature treatment) saves lives. Why take a chance?

In 1922, American doctors formed a new political party called the Medical Rights League. The party aimed to lobby the government for permission to prescribe what substance?

Beer. In 1920, the U.S. Congress passed the Prohibition Enforcement Act, making it illegal to make, sell or buy an alcoholic beverage. This presented a problem for doctors, many of whom were used to recommending a shot or two to their patients for medicinal purposes. Despite the lack of scientific evidence, physicians commonly prescribed brandy or whiskey for conditions ranging from anemia and influenza to indigestion, cancer and heart disease.

In order not to interfere with such "treatments," the law had a provision for medical alcohol. Doctors could apply for a permit to prescribe liquor as long as no patient had access to more than one pint within any ten-day period. This escape clause did not mention beer, and when physicians began to submit requests to prescribe it, claiming among other things that it was an excellent sleep aid and booster of breast milk production, the government had to make a decision. Attorney General Mitchell Palmer, much to the concern of extreme prohibitionists, decided that beer, like liquor, could be prescribed for medical purposes.

Congressman Andrew Volstead, who had drafted the original Prohibition Act, reacted angrily and sponsored a supplementary bill to ban medicinal beer, which he was satisfied to see become law in 1921. This led directly to the formation of a political party, dubbed the Medical Rights League, led by Dr. John Davin, that lobbied strongly for the elimination of the anti-beer law. It isn't that the doctors were so bent on prescribing beer; they were angered by what they saw as the government's attempt to interfere with the practice of medicine. It was they, the doctors maintained, not politicians, who ought to decide what medications their patients should be taking. And if they thought that beer was beneficial as an old-age tonic or as a treatment for dyspepsia, as many did, then they should be allowed to prescribe it.

Doctors mounted a legal challenge against the bill banning the prescribing of beer, and it went all the way to the Supreme Court, where it was defeated by a vote of five to four. The justices, it seems, did not believe in the therapeutic properties of beer. But the whole issue became moot in 1933, when Prohibition was repealed and beer began to flow abundantly.

Whether the beverage really has medicinal benefits is still being debated. While there is no evidence that it increases breast milk production, the barley and yeast used to produce beer do provide some B vitamins, although hardly in significant amounts. Some studies have shown a reduced risk of blood clots due to the anticoagulant effect of alcohol in beer as well as an increase in

HDL, or "good cholesterol," probably due to antioxidants derived from barley. There is also some evidence that drinking moderate amounts of beer (one to two a day) can lower blood pressure. Still, the best reason for the occasional beer is that it tastes good. Unfortunately, beer production is not exactly an environmentally friendly undertaking. If we gave up drinking beer, the barley saved could be used to feed millions of starving people. But does anyone want to propose another Prohibition? I didn't think so.

What substance was originally used to determine the proof of an alcoholic beverage?

Gunpowder. The use of the term "proof" to indicate the alcohol content of a beverage dates back to the sixteenth century. A spirit was then said to be proven to contain a significant amount of alcohol if it ignited in a certain way when mixed with gunpowder. Gunpowder back then was a mixture of sulphur, charcoal and potassium nitrate, better known as saltpetre. If an alcoholic beverage mixed with gunpowder flared up when ignited, it contained too much alcohol. If it just sizzled, there was too little. But if it ignited just the right way, it was a proven spirit.

We still use the term "proof" today, but there is no gunpowder test involved. The proof of a beverage is defined as double the alcohol content. For example, 90-proof gin contains 45 per cent alcohol. That's the strength of gin you need for soaking white raisins. And why would you want to do that? Because according to folklore, eating these raisins can relieve the pain of arthritis. Just take nine of them a day for relief, we're told. If there is no improvement, increase the dose to thirty-six raisins a day. Of course, there is no clinical proof that this works, but I wouldn't be surprised if the 90-proof gin raisins make you feel better.

What substance used in beer production would present a problem for vegans?

Isinglass. Most people like their beer to be absolutely clear and are troubled by a hazy appearance. The haze can be caused by one of two factors. Yeast, which is what makes fermentation possible, tends to settle to the bottom and can eventually be filtered off. But some yeast cells do remain in suspension and cause the beer to be cloudy. This is where isinglass comes in. It is a protein, a form of collagen, that when added to beer acts sort of like a net as it falls to the bottom, ensnaring floating yeast cells. Wines are also sometimes clarified in this fashion.

So where does the vegan concern arise? True vegans do not consume any form of animal product. And isinglass is isolated from the swim bladders of fish, usually the beluga sturgeon. Not all beers are clarified with isinglass. The technique is most often applied to cask-conditioned ales that are consumed directly from the cask where fermentation occurred. But sometimes isinglass is used by brewers to reclaim batches of beer that didn't filter properly.

Yeast haze is not the only issue with beer clarity. Non-microbiological particles such as proteins, polyphenols, carbohydrates and plant fibre can also contribute to a cloudy appearance. Isinglass is not helpful with these, but carrageenan, a carbohydrate found in a type of seaweed known as Irish moss, is. Carrageenan coagulates proteins, and as these sink to the bottom they drag down other insoluble impurities. Both isinglass and carrageenan are referred to as finings, and their mode of action is also called fining. Anyone who has ever attempted to brew beer at home knows that it is virtually impossible to produce a clear beverage without fining.

There is yet another issue with isinglass, and that is the question of whether it can cause a reaction in people who suffer from a fish allergy. Theoretically this is possible, but there is no record in the scientific literature of any case that has been directly linked to isinglass. The amount of this fish protein that actually can be found in the final beverage is trivial.

Soft-drink bottles are made of a plastic called polyethylene glycol terephthalate, or PET. While this plastic is fine for storing soft drinks, it would not be recommended for storing your homemade wine. Why?

While PET has a very low permeability when it comes to carbon dioxide, it readily allows oxygen to pass through. And oxygen is the enemy of wine. When we talk about storing soft drinks, permeability to carbon dioxide is the critical factor. A beverage that loses carbonation loses its appeal. But while oxygen passing into a plastic soft-drink bottle from the air may react with some of the flavour components, the effect would be minor given that we don't store soft drinks for extended periods. But of course we do store wine to age it. And this is where oxygen becomes a problem.

Grape juice contains a variety of compounds called polyphenols, which can react with oxygen and produce a variety of colours and flavours. This really is the same chemistry that occurs when an apple is cut and is exposed to the air. Reaction between polyphenols and oxygen produces the brown discoloration. Not only will the apple slices look different, they will also taste different. The same thing can happen with wine.

White wine is more susceptible to such changes because it lacks some of red wine's coloured compounds, the anthocyanins, which can act as antioxidants. Sulphur dioxide is also an effective antioxidant, which explains why compounds that release this gas, such as sodium bisulphite, are used to preserve wine. Burning sulphur inside wine barrels to produce sulphur dioxide is also an age-old method of preservation.

Now, back to our plastic bottles. As we have seen, empty soft-drink bottles are too permeable to oxygen and are not appropriate for storing wine. But wine can be purchased in plastic containers, although I suspect a true oenophile would look warily upon this method of marketing. So how do the marketers solve the problem of oxygen permeability? By sandwiching a layer of an oxygen-impermeable plastic between layers of food-grade polyethylene or polypropylene. Ethylene-vinyl alcohol copolymer is ideal for this purpose since it allows very little oxygen to pass through—a bit of ingenious chemistry. So while it is not a good idea to store your wine in old soda bottles, it is quite acceptable to purchase wine in plastic containers.

What did a recent study reveal about lemon slices that are commonly placed on the rim of a beverage glass?

The possibility of bacterial contamination. Anne LaGrange Loving, a microbiology professor at a community college in New Jersey, became concerned when she saw a drink decorated with a lemon wedge being served in a restaurant by a waitress with dirty fingernails. Loving was tuned in to possible microbial contamination, having previously carried out a study on bacteria contaminating communion cups.

She now decided to investigate whether lemon wedges were a possible source of bacterial contamination. In a thorough investigation, forty-three restaurant visits were made and seventy-six lemon wedges from twenty-one restaurants were swabbed for microbes, both on the skin and the flesh of the lemons. The restaurant staff were unaware of the experiment. Seventy per cent of the samples were found to be contaminated with any of twenty-five varieties of bacteria or yeast. It came as no surprise that reporters saw a juicy story here and featured articles about yet another hazard in our daily life. It seems we couldn't even trust those supposedly healthy lemon wedges that decorated our drinks!

Well, let's not panic about the results of this study. The fact is that microbes are everywhere; you can swab practically any surface anywhere and isolate some bacteria or yeast. If we want to evaluate the health consequences, we need to know which bacteria, and in what numbers they are present. There are thousands of different microbes with which we happily coexist. Relatively few cause disease. True, young children or people who are immuno-compromised are more at risk from disease-causing microbes. And some of these were found on the lemon wedges. But in this case there was no quantitative analysis at all, so we have no idea how extensive the contamination was. Just because the microbes are present does not mean they are a risk.

The scientific literature has no record of any food poisoning outbreak traced to lemon wedges. The fact is that we ingest all kinds of bacteria every day. Our food supply is certainly not sterile, and does not need to be. Of course, the prospect of a waitress placing a lemon wedge on a drink with hands that have not been washed after a trip to the bathroom is not an attractive one. But the bacteria on the lemon wedges could also have come from the hands of pickers—or indeed from anyone who handled the fruit on its long journey to the beverage glass. Sure, restaurant workers should use gloves or tongs to handle lemon wedges, but frankly, I'd be more concerned about what is in the glass than what is on it. With all the research emerging about the consequences of excessive sugar intake, it may be healthier to suck on the lemon instead of drinking the beverage. That is, of course, assuming the lemon wedge has been properly handled.

 # CONSUMER
ISSUES

If you were told to improve your sulphoraphane levels, what would be your best plan of action?

Eat more broccoli and cabbage. Sulphoraphane is one of the most potent anti-cancer compounds ever isolated from a natural source. When rodents are fed sulphoraphane and then are exposed to a cancer-causing substance, they are less likely to develop tumours. The best sources of sulphoraphane are broccoli and cabbage. They may give you gas, but they also gas up your immune system.

Butylated hydroxytoluene (BHT) is often seen listed as an ingredient in cereals. What undesirable reaction is it trying to prevent?

Reaction of fats in the cereal with oxygen from the air. One of the problems with storing foods is that their fat content can go rancid.

While this is not a health hazard, it is certainly undesirable in terms of smell and taste. Rancidity refers to the decomposition of fats into volatile compounds, known as aldehydes and ketones, which can significantly affect scent and flavour. This reaction is initiated by oxygen.

Different fats have different propensities for reacting with oxygen. It all depends on their molecular structure. Unsaturated fats, characterized by having carbon-carbon double bonds in their structure, react readily with oxygen, producing intermediates called free radicals that quickly decompose to the unsavoury aldehydes and ketones. Saturated fats, the ones implicated in the buildup of fatty deposits in arteries, are not prone to reaction with oxygen. Exclusion of oxygen is one way of preventing rancidity but is not practical. However, additives are available that can neutralize free radicals before they have a chance to give rise to the compounds that characterize rancidity. These are the famous antioxidants, with BHT being a prime example.

BHT—unlike vitamins C and E, which can also act as antioxidants—does not occur in nature. It is a synthetic compound whose molecular structure was designed to make it fat soluble and to allow it to react with free radicals. Like any other food additive, BHT had to undergo extensive animal testing before it was allowed on the market. But in spite of such testing, it has not been able to avoid controversy. Alarmists point to some studies in which an increased incidence of cancer was noted in rodents fed large amounts of BHT.

The literature, however, also records studies showing a decreased risk of cancer. It just goes to show that one can dredge the scientific literature these days and find studies to support virtually any point of view. Curiously, some of the opponents of BHT as an additive see no problem with selling capsules of BHT in health food stores as a dietary antioxidant supplement. The dose in those supplements is thousands of times greater than the amount of BHT used as a food additive. There is absolutely no evidence that BHT added to any food presents a health risk. True, it isn't an essential additive in cereals. If the product is not going to be stored for very long, there is no great risk of rancidity. But manufacturers look for long shelf lives, in which case BHT can help maintain freshness.

Why are farmers who grow cacao trees scared of the witch's broom?

Witch's broom is a fungus that just might sweep away the farmers' livelihood. Most of the world's chocolate is made from the fermented and roasted seeds of cacao trees grown on small farms in Africa, but sadly, diseases currently destroy about a third of the crop every year. This could get far worse if witch's broom—or frosty pod, another fungus—invades the plantations.

In the 1980s, witch's broom devastated Brazil's cacao trees, and agronomists believe it is only a matter of time until such nasty fungi find their way into plantations in Ghana and the Ivory Coast. That's why plant scientists are studying varieties of wild cacao trees, hoping to find species that are high yielding, drought tolerant and resistant to diseases. The idea is to examine the DNA of trees that have such properties and identify the genes responsible for these traits. Then new varieties of cacao trees can be bred, and those that have the desired genes can be immediately cultivated instead of having to wait years until a tree is fully grown to evaluate its properties. Huge corporations such as Mars are at the forefront of such research because their profits depend on the availability of cacao beans— you can't make chocolate if you don't have the raw material. And these days, chocolate is hot. "A chocolate bar a day will keep the doctor away" is beginning to replace the more common "apple a day" slogan.

Marketers are capitalizing on the high antioxidant content of cacao powder and have begun to promote chocolate as a quasi drug. They are stumbling over each other in their haste to become the greenest and healthiest purveyors of chocolate. Instead of taste, they talk of oxygen radical absorption capacities, organic raw ingredients and handcrafting. Some, like Xoçai, a multilevel-marketed product, will toss in omega-3 fats, lactobacillus and bifidobacteria or powdered acai berries and blueberries. The company touts the high antioxidant potential of its product and suggests eating three squares a day for optimal effect—not to mention for optimal profits. Consumers will have to shell out about a hundred dollars a month. Put that money toward fruits and vegetables and you'll get a lot more antioxidants.

The makers of Tru chocolate go even further. They throw in everything but the kitchen sink: various extracts of green tea, black pepper, kidney beans, oranges, lemons and noni fruit. Whether the small amounts of these ingredients do anything is highly debatable. Then there is Gnosis chocolate, created by a "certified holistic health counselor." Gee, could we possibly ask for more expertise? You'll be comforted to know that she has worked hard to select the most potent combinations of such superfoods as agave nectar and crystal manna, a type of algae, to add to her handmade confections. There is another ingredient that is listed on every variety: love. I wonder where that grows. But I'm happy to hear that Gnosis is made with "green integrity" and "Gaia consciousness." I think I can do without the Gaia consciousness—and all the other overhyped additives as well.

Any dark chocolate bar that has about 70 per cent cocoa powder is good enough for me, although I do admit to favouring a Dove bar. In fact, there is even clinical evidence that, eaten every day, the flavonoids it contains, particularly epicatechin, improve the way that blood vessels dilate and relax. The vessels lose stiffness, which is a marker for heart disease. At least so far, nobody has tried to tout the benefits of chocolate-covered doughnuts. It will come, though. Probably fortified with blue-green algae, wheat grass juice or red wine extract.

If you were appertizing, what would you be doing?

Canning food. And you would be doing it in a fashion similar to that introduced by French inventor Nicolas Appert back in 1810. His motive for the invention? The usual one: money! Appert was after the twelve thousand francs the French government was offering to anyone who could come up with a practical method for preserving food.

Supplying food for armies on the march was an age-old problem, particularly in winter, a fact recognized by Napoleon. "An army marches on its stomach," the emperor was fond of saying. And it seems that stomachs did not march all that well on the dried, smoked, pickled or salted foods available at the time. A process that would make more pleasing meals available to Napoleon's marauding soldiers was sorely needed. Appert, trained as a cook and confectioner, took up the challenge.

Results did not come overnight. It took Appert about fourteen years to work out a method of preservation that involved sealing food in glass jars, followed by immersion in boiling water. Just what prompted this particular line of investigation is not clear.

That wine could be preserved by adding boiled must (newly pressed grapes) was known since antiquity, but it is also possible that Appert was intrigued by Italian biologist Lazzaro Spallanzani's famous experiment. In 1765, Spallanzani attacked the widely held theory that life could be spontaneously generated. Beef broth, for example, if allowed to stand for a few days, would produce cavorting "animalcules" that could be readily observed through a microscope. Animalcules are now known as microbes.

Where did they come from? Since they could not be seen in the original broth, went the prevailing wisdom, they had to have been "spontaneously generated." Spallanzani was skeptical. He showed that broth that had been boiled and then sealed did not give rise to the animalcules. Today, of course, we understand why: any microbes originally present in the broth were killed by the heat treatment, and sealing the jars prevented any further contamination. Spallanzani did not offer any such explanation, but did argue that his experiment was evidence against the spontaneous generation of life. But supporters of the theory discounted his work, choosing to believe that Spallanzani's treatment excluded oxygen, which they thought was vital to spontaneous generation.

Since theories of spontaneous generation were widely discussed at the time, it is quite possible that Appert capitalized on Spallanzani's observation that boiled beef broth did not give rise to microbes. Admittedly, this is speculation, because there is no evidence at all that Appert ever proposed any explanation as to why his process of preserving foods worked. But work it did, and in 1810 he collected the prize money that had been offered.

Some fifty years later, Louis Pasteur would explain that food spoilage was caused by microbial activity, and that this could be prevented through pasteurization—which, of course, was just a version of appertizaton.

With his prize money, Appert set up what we would call the first commercial cannery in Europe, although he used glass jars and not cans. The metal can was introduced soon after, by Peter Durand, since metal was cheaper and more durable than glass. Amazingly, the can opener was not invented until 1858. So how did soldiers open their canned foods? Well, bayonets came in handy. And on an empty stomach, probably not much thought was given to whatever else those bayonets might have been plunged into.

One of the dishes served at an avant-garde New York restaurant, wd-50, is based on noodles made from shrimp, without the use of any flour. What chemical does chef Wylie Dufresne use to create this improbable delicacy?

Transglutaminase, an enzyme with the ability to link protein molecules. Enzymes themselves are special protein molecules that plants and animals use to carry out a variety of biological functions. Our bodies produce transglutaminase to aid in blood clotting, a process that requires protein molecules to form interlinked complex structures. Skin and hair are also composed of proteins that have been bound together, and transglutaminase plays a role here too.

In the 1990s, the food industry discovered that this enzyme can be isolated in good yield from the bacterium *Streptoverticillium mobaraense* and that it can be used to "restructure" meat, fish and poultry. For example, with the help of transglutaminase, bits of chicken left over after the carcass has been processed, instead of being discarded as waste, could be "glued" together to produce—you guessed it— Chicken McNuggets. Similarly, artificial crab legs and shrimp could be made by sticking together ground pieces of cheaper seafoods such as pollock.

While the taste of such artificial foods can be criticized, there is no health issue associated with the use of transglutaminase. As any other protein, it is readily broken down into its component amino acids in the digestive tract. Transglutaminase is produced for the food industry under the name Activa by the giant Japanese company Ajinomoto, which also markets monosodium glutamate.

Transglutaminase did its work quietly behind the scenes until celebrity chef Heston Blumenthal brought it out of the shadows at the Fat Duck, the restaurant on the outskirts of London that has been labelled by many as the best in the world. Blumenthal's enthusiasm for creating novel dishes with the enzyme rubbed off on Wylie Dufresne, who managed to grind shrimp into noodles with the help of transglutaminase and served it on a bed of smoked yogurt.

Chefs who pursue what has been called molecular gastronomy—essentially, the application of scientific principles to the creation of new dishes—are exploring the possibilities that transglutaminase may afford. Around the corner are filet mignon with strips of bacon glued to its surface, fish with chicken skin stuck to it to enhance flavour, and shrimp burgers held together by cross-linked proteins. And how about steaks made from leftover meat scraps? Yummm! Egg whites are also made of protein. So it seems to me that a hardboiled egg can be cut in half, the yolk removed and something inserted in its place before the two halves are glued back together with a sprinkling of transglutaminase. I'm sure there is a magic trick hiding in there somewhere.

You've had a delicious serving of chili that was slow-cooked in a Crock-Pot. A couple of hours later, you begin to experience nausea and vomiting, followed by diarrhea. What is the most likely cause of the problem?

Symptoms of nausea, vomiting and diarrhea are often suggestive of bacterial contamination; however, they generally take at least six hours to appear. In this case, reaction is far more likely due to the presence of a class of proteins known as lectins. These proteins are found in a large variety of plants and animals, where they serve a variety of biological functions based on their ability to bind to specific carbohydrates. In animals, these functions include helping cells adhere to each other and marking foreign substances for removal from the bloodstream by the liver. The function of lectins in plants is less clear, but the fact that some—such as phaseolin and arcelin—have insecticidal properties suggests that these compounds help protect the plants from their insect enemies.

The highest concentration of lectins is found in plant seeds, a possible evolutionary result of the plant trying to stop animals from eating its seeds by triggering adverse symptoms characteristic of lectin ingestion. Since we are also animals, we can also get sick from eating plant lectins. But luckily, these proteins are destroyed by heat, so we are only at risk if we eat uncooked plant material that has a high concentration of lectins. Kidney beans are a prime example.

As few as four or five raw beans can cause misery. But boiling in water for at least ten minutes destroys most of the lectins. However, in slow cookers the temperature may not rise above 60°C, and lectins can survive intact. Even worse, heating to 80°C actually increases the concentration of lectins. Canned beans are fine because they have been previously heated to a high temperature, but some people just soak kidney beans to soften them before using them in a salad. This is dangerous and should not be done.

Certainly, the presence of lectins should not scare anyone away from eating beans, which have a variety of health benefits. Beans contain both soluble and insoluble fibre, meaning that they can reduce blood cholesterol, prevent blood sugar levels from rising too rapidly after a meal and speed the passage of fecal matter through the intestinal tract. Kidney beans are fat free and are a good source of iron and molybdenum, a trace mineral required by some enzymes. Of course, there is one other side effect of bean consumption: copious gas production. But if you cook your beans properly, that's the only side effect you'll have to put up with.

L-cysteine is an amino acid used in some bakery products as a dough conditioner. What raw material used for the production of this chemical has caused controversy?

Human hair. Hair is essentially composed of proteins, which in turn are made up of amino acids joined together in a chain. L-cysteine is one of these amino acids. The chain can be broken down using various chemical treatments to liberate the amino acids, which can then be separated and purified. L-cysteine obtained in this fashion has a number of industrial uses.

As a dough conditioner, it reduces viscosity and makes the dough softer and easier to work. Dough that has been treated with L-cysteine also increases its volume more effectively during baking. All of this happens because cysteine disrupts the structure of gluten, that network of protein molecules that determines the consistency of baked goods. Specifically, L-cysteine can break apart the disulphide bridges linking adjacent protein molecules together in gluten, thereby weakening the "scaffolding" that lends consistency to the dough.

The food industry also uses L-cysteine as a flavour additive because of its ability to react with sugars to yield meat-like flavours. Even the pharmaceutical industry gets into the act. L-cysteine is needed to produce N-acetylcysteine for the treatment of acetaminophen overdose, as well as for reducing the thickness of mucus in conditions such as bronchitis and emphysema. Here, too, it is the ability of L-cysteine to break disulphide bonds that is of importance, as it is in some permanent-wave lotions used by the cosmetics industry. When L-cysteine disrupts its disulphide bonds, hair can be shaped using rollers, and the disulphide bonds are subsequently reformed by the application of hydrogen peroxide.

As far as the activity of the L-cysteine is concerned, the source from which it is produced is irrelevant. Whether it is obtained from duck feathers, pig bristles or, traditionally, from human hair collected from barbershops in China, does not matter. There is no way to distinguish an L-cysteine molecule that originated in the feather of a duck from one that was sourced from hair. But the origin does make a difference to some people who for religious or ethical reasons wish to avoid any product that stems from a human source. Alternatives are available.

The Japanese company Ajinomoto produces L-cysteine by chemical synthesis, and the German company Wacker uses a fermentation process to make the compound from corn. L-cysteine produced by either of these methods is suitable for vegetarian, kosher or halal food preparation. All baked goods that have been processed with L-cysteine must list the chemical as an ingredient, but not its source. In Europe, the Food Standards Agency specifies that only L-cysteine produced from duck or chicken feathers or from pig bristles may be used. In North America, the only way to determine the source is to track down the producer and ask. A hairy issue indeed.

When Hernán Cortés visited Mexico in the sixteenth century, he discovered that the natives tenderized meat by wrapping it in the leaves of a tree. What tree?

Papaya. The leaves and fruit of the papaya tree produce papain, an enzyme that breaks down protein molecules. Since the fibres that toughen meat are made of proteins, wrapping meat in papaya leaves will lead to tenderization. Papain can be extracted from papaya and is available commercially as a white powder for cooks to sprinkle on meat. It is also included in some contact lens–cleaning solutions due to its ability to digest proteins—in this case, those that deposit on the contact lens from the eye.

Papain supplements are also sometimes sold as digestive aids, and in theory should be effective in helping to break down proteins in the digestive tract. Just how well orally ingested papain stands up to the acidic environment of the stomach is questionable. Treatment of bee or jellyfish stings with papain is not unreasonable, since we know that the toxins injected are proteins. But you certainly want to keep papaya away from your Jell-O! Papain will break down the proteins that make up gelatin and you will be left with a liquidy mess.

While we know what papain does, we don't really know why it does it. Why should a fruit tree produce protein-destroying enzymes? Trees reproduce by dispersing seeds, and an effective way to do this is by attracting animals to eat the fruit. One theory about papain is that its protein-digesting ability helps rid animals of intestinal parasites. This is certainly a possibility since studies have shown that papain can destroy live tapeworms. Indeed, in some tropical areas where intestinal parasites are endemic, natives use the latex from papaya as a drug, with apparent success.

Why is sodium acid pyrophosphate commonly added to commercial french fries?

To prevent the potatoes from forming black spots. Sodium acid pyrophosphate is referred to as a sequestering agent, which simply means it has the ability to bind with metal ions. In this instance, the relevant metal is iron, which occurs naturally in potatoes. We aren't talking about miniature nuts and bolts embedded in potatoes, we are talking about ions of iron. An ion is an atom that has gained or lost electrons. When an atom of iron loses two electrons, it becomes a doubly charged positive ion, known as a ferrous ion. This is not a problem as far as discoloration goes, but when it reacts with oxygen in the air, it loses another electron to oxygen and becomes the triple positively charged ferric ion. This is when the potato can become discoloured.

Ferric ions react with chlorogenic acid, another naturally occurring compound in potatoes to form a dark complex. While this is not a health issue, the dark spots are unappetizing. Sodium acid pyrophosphate prevents this from happening because it sequesters, or binds, the ferric ions. Potatoes used by the fast food industry are usually frozen and can be stored for a long time, allowing ferrous ions to oxidize. That's why they are treated with sodium acid pyrophosphate before being frozen.

Other methods are also available. Purified, spray-dried gum acacia and co-processed gum acacia and gelatin in combination with calcium chloride show considerable promise. Calcium ions compete with ferrous ions and thereby reduce the formation of the oxidizable ferrous ion–chlorogenic acid complex. And if you should see your potatoes developing black spots when you cook them at home, just add a touch of vinegar to the cooking water. The acid destroys the ferric ion–chlorogenic acid complex.

What is the oldest known method of food preservation?

The use of wood smoke. The earliest deliberate application of chemistry was probably in cooking. Primitive man—or perhaps woman—learned that heating meat made it taste better and keep longer. And somewhere along the way, he or she found, probably as a result of trying to dry meat over a fire, that smoke improved its flavour and its keeping qualities.

Wood consists essentially of carbohydrates in the cellulose family, which make up the cell walls, and lignin, a substance that binds the cells together. When these are heated, they produce a variety of flavourful breakdown products. They also yield formaldehyde and acetic acid, which are established preservatives. Lignin also burns to produce phenols, which are toxic to the bacteria and fungi that cause food to spoil. Many of these phenols are also antioxidants, meaning they can slow down the development of rancid flavours that result from the reaction of fats with oxygen.

Unfortunately, wood smoke also contains compounds that are known carcinogens. Pyrene, fluoranthene, benzpyrene and benzanthracene can all cause cancer in test animals. That does not necessarily mean that the small doses humans are exposed to do the same, but the possibility does exist. There is, for example, a high incidence of stomach cancer among Icelanders and Baltic fishermen who consume a lot of smoked fish.

The extent to which carcinogens form in the food depends on the kind of wood used and the temperature to which it is heated. Mesquite, for example, which burns at a very high temperature, produces twice the amount of carcinogens as hickory. One way of reducing the risk of smoked foods is through the use of "smoke solutions." These are made by heating sawdust and passing the resulting smoke into water. The temperature can be controlled to produce fewer carcinogens, and in any case these are insoluble in water. Most of the compounds responsible for smoke flavour will dissolve. Treating meat with such a solution instead of hanging it in a smokehouse results in a safer product. In any case, smoked foods also tend to be high in fat and salt, so it is a good idea to eat them infrequently.

What common use links the following substances: caramel, annatto, cochineal, betanin, turmeric, saffron, paprika and elderberry juice?

All of these are used as "natural" food dyes. Food dyes are becoming increasingly more controversial as public concern rises about exposure to chemical additives in our food supply. Since, unlike preservatives or flavours, food dyes serve only a cosmetic function, any risk associated with their use has to be very strictly evaluated.

The majority of dyes used in the food industry are synthetics made from chemicals derived from petroleum. Over the years, some of these have been withdrawn from use when animal studies suggested possible harm. Perhaps the most famous case was the United States' 1976 withdrawal of Red Dye #2, also known as amaranth, based on some studies that suggested it could cause cancer when fed in high doses to female rodents. The food industry complained that this was not a rational ban, since a human would have to drink some 7,500 cans of a beverage dyed with Red #2 every day to reach the rats level of consumption. Canada agreed and did not delist the use of this dye. Which just goes to show that different regulatory agencies can come to different conclusions based on the same evidence, meaning that the evidence is not conclusive. Some studies have also linked Red Dye #3, known as erythrosine, with thyroid cancer, but again the doses in these studies were far higher than possible human exposure, and the dye is still approved for use in foods in Canada, Europe and the U.S.

While the cancer connection is insignificant, food dyes can cause other problems. Tartrazine, a yellow dye, can cause hives, but this happens in less than 0.01 per cent of those exposed. But the issue that has stirred up the most controversy is the possible link between food dyes and attention deficit disorder in children.

A well-designed study published in the British medical journal *The Lancet* in 2007 concluded that certain food dyes, as well as the preservative sodium benzoate, exacerbated hyperkinetic behaviour in some children who were predisposed to such behaviour. Certainly, the effect was not overwhelming, with dyes being responsible for altered behaviour in roughly 10 per cent of children. Still, one can make the valid point that food colourants are unnecessary, and their use can be readily reduced or in some cases eliminated. A case in point is a McDonald's strawberry sundae. In the U.K., where public opinion is forcing the elimination of dyes, the sundae is made with colour from real strawberries, whereas in North America, Red Dye #40 is used. Natural dyes such as betanin from beets or carotenoids from the seeds of the annatto tree are, in general, less likely to cause problems, both in terms of health effects and marketing. But reactions are still possible. Allergic reactions to cochineal, a red colourant extracted from the female cochineal insect, are well documented.

A Turkish company is set to market foods supplemented with human lactoferrin, a protein that has immune system–enhancing properties. Where will it obtain the substance?

From the milk of cows that have had the human gene that codes for the production of lactoferrin introduced into their DNA. Lactoferrin, found in human saliva, tears and breast milk, has been shown to have anti-infective and anti-inflammatory properties. In fact, lactoferrin is one of the components of breast milk that protects babies from infections and promotes lifelong healthy immune function. There is evidence to suggest that lactoferrin as a dietary supplement may be helpful in the treatment of peptic ulcers caused by *Helicobacter pylori* infections, in the treatment of inflammatory conditions such as Crohn's disease, and in the treatment of viral diseases like hepatitis.

Cows also produce lactoferrin in their milk, and most investigations of the potential therapeutic effects of the protein have used bovine lactoferrin. While similar to the human version, it is not identical. Indeed, human lactoferrin should be more effective as a dietary supplement but is more difficult to obtain. A Dutch biotech company, cleverly named Pharming, has solved that problem by inserting into a cow's genome the human gene that codes for the production of lactoferrin.

Basically, the technology involves isolating from human DNA the gene that codes for lactoferrin and inserting it into the DNA in the nucleus of a cell taken from a cow. This cell is then fused with an egg cell from which the nucleus has been removed, and the new genetically modified cell is then implanted in a cow's uterus. If all goes well, she will give birth to a calf that will produce human lactoferrin in her milk. This may sound like some sort of monstrous interference with nature, but the transgenic cows being raised on a farm in Wisconsin are perfectly normal, and the lactoferrin they produce in their milk is identical to the version humans produce.

Whether or not this protein, if added to foods or taken in pill form, will produce significant health benefits remains to be seen. Experiments in animals have shown that lactoferrin can be absorbed into the bloodstream from the digestive tract, so physiological effects are possible. Aslan, the Turkish company that has signed an agreement to commercialize lactoferrin in the Middle East and Russia, is betting that there is a market for this dietary supplement, even though there is no clear-cut evidence of efficacy or of the dosage that may be required to improve immune function. Perhaps the best hope for the future is to add human lactoferrin to baby formula, given that in North America only about 11 per cent of babies are exclusively breast-fed for the first six months of life.

Cow's milk, the source of infant formula, provides the (less effective) bovine version of lactoferrin. Addition of human lactoferrin to rehydration formulas used to treat diarrhea, an affliction that kills close to two million children a year, also has great potential. And as a final kicker, researchers investigating the use of lactoferrin to fight gum infections stumbled on what may be the most marketable aspect of this protein. The participants in the study lost a significant amount of abdominal fat, the type of fat that has been linked to heart disease and diabetes. This seems to mesh with other research that has noted that breast-fed babies are less likely to become obese or develop diabetes.

Why should potatoes be stored in the dark?

Light stimulates them to sprout and produce compounds called glycoalkaloids, of which solanine is the most abundant. These compounds are toxic to pests, so the potato produces them to protect itself during the sensitive growth stage. In a high enough dose they are toxic to humans as well. A sign that the potato has been improperly stored is the appearance of green discoloration. The green colour itself is due to chlorophyll and is not toxic, but its appearance signals that solanine is being formed. Eating lots of green potatoes can cause itchy skin around the neck, drowsiness, vomiting and, rarely, rapid heartbeat. Glycoalkaloids are probably the most widely consumed natural toxins in North America. But don't worry, they are still consumed in amounts too small to cause potato poisoning. One can never repeat the Paracelsian doctrine often enough: only the dose makes the poison!

The famous Fat Duck restaurant in England has fitted one of the taps in its kitchen with a water-softening filter. What for?

Cooking vegetables. Everyone knows that vegetables are good for us, but cooking them presents a challenge. We want the nutrients to be retained, which means that we want a relatively short cooking time, but a short cooking time may leave us with a tough texture.

The crispy structure of vegetables is due to complex carbohydrates called pectins. During cooking, pectins become soluble and are extracted into the cooking water, leaving us with softened veggies. But if the cooking water is "hard," meaning that it has a high calcium content, a problem arises: calcium ions form cross-links between pectin molecules, which makes them less soluble, keeping the vegetables tough. This means that a longer cooking time is needed, which thereby reduces nutrients and affects the colour of the vegetables in an undesirable fashion. The longer the cooking time, the more likely that the green colour of chlorophyll is lost.

The chlorophyll molecule consists of a large ring of atoms with a magnesium ion sitting in the middle. When the magnesium ion is lost, as happens during prolonged cooking in water, green chlorophyll changes into brownish pheophytin, the colour of overcooked vegetables. A water-softening filter removes calcium ions, which means that vegetables can be cooked more quickly to the desired texture without sacrificing colour or nutrients.

The chemistry here can be readily confirmed by a simple experiment in which dried peas are soaked overnight in tap water, in softened water, and in water to which some calcium chloride has been added. The drained peas can then be cooked in simmering water and periodically sampled to evaluate texture. Peas cooked in soft water will become tender very quickly, while those cooked in hard water will take a longer time to soften. If avoiding softening is desirable, as in the case of pickles, then calcium chloride can be added to the pickling mixture.

Why would someone swallow polydimethylsiloxane?

Because they hope to reduce the discomfort caused by excess gas in the gastrointestinal tract. Polydimethylsiloxane, commonly called simethicone, is a defoaming agent. A foam consists of bubbles that are just pockets of gas surrounded by a thin film of liquid. The nature of the liquid film determines foaming ability. Water doesn't foam because its molecules have a strong affinity for each other, making it difficult to produce a thin film that will stretch. But surfactants, such as molecules of soap, get in between water molecules, reducing their attraction for each other. Bubbles can then form. Anti-foaming agents also interfere with the affinity of molecules for each other in a liquid, this time to an extent that the interaction is so weakened that the thin layer of liquid surrounding the bubble bursts. The immediate result is that small gas bubbles coalesce into bigger ones, which then readily burst, releasing their contents. If the collapsing bubbles are in the intestine, the escaping gas declares its presence as a burp.

Intestinal gas can have several origins. Swallowed air, bacterial fermentation in the colon, lactose intolerance and irritable bowel syndrome can all result in bloating. Simethicone, in theory at least, should help by bursting bubbles, allowing the gases to be released. Satisfaction with the use of simethicone to relieve intestinal gas is mixed, but there is no doubt that such substances, collectively termed silicones, are very effective industrial defoaming agents.

Foam can be a huge problem in boilers, in paint manufacture, paper production, waste-water treatment and numerous other industrial processes. Silicone-based defoamers solve the problem. They don't, however, solve the problem of bloated claims about the toxicity of this substance. In 2006, an article began to make the rounds on the Internet, claiming that a company selling cheese to Pizza Hut was illegally using polydimethylsiloxane in the cheese to prevent it from bubbling up during baking. The explanation offered was that the pizza cheese has starch added to act as an extender, which is true, and that cooking water and starch results in bubbles, also true.

A silly analogy was then drawn to wallpaper paste, which is also composed of starch and water. The implied message was that Pizza Hut's pizza was topped with wallpaper paste and an industrial anti-foaming agent. It seems the writer of the article looked at an old patent for pizza cheese that indeed referred to the possible use of silicone antifoaming agents in cheese. But patents are designed to cover a broad range of possibilities, and there is no implication that they will all be used. Indeed, had the writer of the article checked with the cheese producer, he would have learned that polysiloxanes were not being used. Furthermore, even if they were, there would be no problem because polydimethylsiloxane is approved for use in foods ranging from beer and batters to canned fruit and chewing gum. That isn't surprising given that the substance is approved for use, in much greater amounts, in over-the-counter anti-gas medications.

COLOUR SUPPLEMENT

What pigment derives its name from the Latin for "beyond the sea?"

Ultramarine. The stunning blue pigment is not named after the colour of the sea but derives its name from the fact that it was imported into Europe from Asia by sea. The source of ultramarine is the mineral lapis lazuli, the greatest deposits of which are found in the Badakhshan province of Afghanistan. It isn't surprising, then, that the earliest appearance of this pigment is in cave paintings in Afghanistan, dating back to the sixth century. But long before the pigment was produced from the mineral, lapis lazuli itself was used to make jewellery and talismans.

The mines in Afghanistan are believed to have been worked for more than six thousand years. The ground stone was even used as a cosmetic. Cleopatra's stunning blue eye makeup was concocted of ground lapis lazuli. The Romans actually ate the stuff, believing it to be a powerful aphrodisiac. Ultramarine, as a pigment for painting, seems to have been introduced into Europe sometime in the twelfth century but was not widely used because of the expense. Indeed, it was more costly than gold! Manufacturing the pigment from lapis lazuli was a complicated process, involving grinding with melted wax, wrapping in a cloth and then kneading in a dilute lye solution. The chemical makeup of ultramarine is also complex, with sulphur atoms being trapped in a lattice of sodium aluminum silicate.

Perhaps the most famous use of ultramarine was by Michelangelo, who believed that it was the only colour that would do for the brilliant blue of the sky in the Sistine Chapel. Pope Julius II, under whose patronage Michelangelo painted the ceiling with its famous centrepiece of *The Last Judgement*, could well afford the expensive pigment. Other patrons, it seems, could not. Michelangelo's *Entombment of Christ*, which now hangs in London's National Gallery and predates his Sistine Chapel work, is unfinished. A corner of the painting where the Virgin Mary was to be portrayed is blank, apparently because Michelangelo's patron could not come through with a shipment of ultramarine. The Queen of the Sky had to have a dress that reflected her majesty!

Since ultramarine was rare and expensive, artificial methods for its production were constantly sought, but it was not until 1828 that Christian Gmelin, a German professor of chemistry, came up with a synthetic version. It was not a straightforward business. Gmelin mixed a specific type of clay with sodium sulphate, sodium carbonate and charcoal, added some powdered sulphur, then baked the concoction in a kiln. Grinding and washing yielded synthetic ultramarine, although not quite as vivid a blue as the natural version. Apparently the size of the pigment particles differs from those in the natural version, affecting the way light is diffused.

A problem with any ultramarine is that traces of acid will liberate the sulphur content as hydrogen sulphide and destroy the blue colour. So with acid vapours in the air, due both to industrial and natural processes, works of art painted with ultramarine are destined to deteriorate. But you don't have to rush to see the marvellous blue of the Sistine Chapel—we're talking centuries, not days.

Why do baked potatoes not turn green?

The green discoloration is due to the production of chlorophyll when the tuber is exposed to light. Light triggers a process whereby, with the aid of enzymes, naturally occurring compounds in the potato are converted to chlorophyll. Heat destroys these enzymes, and therefore a baked potato cannot turn green. The green colour is unsightly, but in itself does not pose any kind of health threat. After all, we eat lots of chlorophyll whenever we eat green vegetables.

That doesn't mean there is no health concern associated with eating green potatoes. When a potato turns green, another compound called solanine also forms, independent of chlorophyll production. This is potentially troublesome. The theory is that potatoes evolved to produce solanine to ward off fungi at the most critical stage of their growth, which is when they sprout. Solanine has a bitter taste and can cause stomach problems, but a lot of green potatoes would have to be consumed to note this effect. Storing potatoes out of the light prevents discoloration, which is why dark paper or dark plastic bags are ideal. Bags with windows made of green cellophane are fine too, because the wavelengths of light that cause chlorophyll formation are filtered out.

In 1992, Mattel introduced Hollywood Hair Barbie, a doll that came equipped with a bottle of "Magic Hair Mist." Barbie's hair turned pink when sprayed with

the mist, then slowly changed back to the original
blonde. What made this possible?

Phenolphthalein, the classic acid-base indicator familiar from high
school chemistry labs. This chemical is pink in a solution that has a
pH higher than 8.2—in other words, in a basic solution. If the solution is made more acidic, the pink colour disappears. In the case of
Barbie, the hair was treated with an acidic solution of phenolphthalein and allowed to dry. When sprayed with the Magic Mist, a dilute
solution of sodium hydroxide, the hair turned pink. Slowly the pink
colour was lost as carbon dioxide in the air dissolved in the moisture
on the hair, forming carbonic acid, which neutralized the sodium
hydroxide. It was a neat toy—too bad it is no longer available.

Phenolphthalein is also used by crime scene investigators to
detect the presence of blood. The test is based on the ability of
hemoglobin, the oxygen-carrying molecule in blood, to catalyze
the decomposition of hydrogen peroxide into oxygen and water.
Oxygen production is the key to the blood test because it is needed
to produce the form of phenolphthalein that is colour sensitive to
pH. The reagent used by crime scene investigators, known as the
Kastle-Meyer reagent, contains sodium hydroxide as well as the so-
called "reduced" version of phenolphthalein, which is made by
treating phenolphthalein with zinc. This form of the chemical is
colourless at any pH. The reagent is added to a suspicious stain,
followed by the addition of a drop of hydrogen peroxide. Even the
slightest trace of hemoglobin will release oxygen from hydrogen
peroxide, which then converts the reduced phenolphthalein into its
usual pH-sensitive form. Since the solution already contains a base,
sodium hydroxide, it will turn pink. That means the stain could be
blood. It does not absolutely prove this, because there are other
substances that can release oxygen from hydrogen peroxide. Traces
of metals such as copper can do it, as well as some enzymes present
in fruits or vegetables.

And a criminal with some chemical knowledge can fool the test. A number of stain removal products on the market, such as OxiClean, contain sodium percarbonate, which releases oxygen when dissolved in water. Oxygen acts as an effective bleaching agent, removing many stains. But when it is used to launder a bloodstain, it reacts with hemoglobin in such a way as to negate its ability to release oxygen from hydrogen peroxide after the application of the Kastle-Meyer reagent. Usually, though, criminals do not have the time to carry out an elaborate cleanup, and the phenolphthalein reagent serves as an effective method for determining if a stain is blood or not.

"Red ruby" glass first produced in the late seventeenth century is a collector's item. What substance was responsible for the stunning colour?

Gold. Just mention gold, and visions of the lustrous yellow metal spring to mind. But gold isn't always yellow—it can be a brilliant red. It all depends on particle size. Large samples of the metal reflect mostly yellow light, but when particles are extremely small, in the nanometre range, a different phenomenon occurs.

White light is composed of all the colours of the rainbow, each colour characterized by a certain wavelength. Nano-sized gold particles scatter all wavelengths except red, which they transmit. So a sample of water or glass that harbours such tiny particles will appear red. European painters of the fourteenth century seem to have been aware of this unusual property of gold, but it was first documented in detail in 1685 by Andreas Cassius, a Hamburg physician.

Cassius was familiar with aqua regia, the only solvent capable of dissolving metallic gold. This mixture of nitric and hydrochloric acids dissolves gold to yield a yellow solution. But when a second solution, made by dissolving tin filings in aqua regia, is added to the gold solution, a precipitate ranging in colour from purple to pink is produced. Such a precipitate came to be known as purple of Cassius. Upon heating, purple of Cassius yielded a brilliant ruby-red colour. At the time, the chemistry was not understood, but today we know that gold reacts with aqua regia to form soluble gold chloride, which upon reaction with tin compounds yields metallic gold particles. The exact size of the gold particles determines their colour and is a function of how the gold and tin solutions are mixed.

Johannes Kunckel, a German glassmaker, was the first to add purple of Cassius to glass, and he produced samples of stunning beauty. Kunckel, though, may have just reinvented the wheel. One of the numerous amazing items on display at the British Museum is the Lycurgus Cup, probably made in Rome in the fourth century AD. The cup looks green in reflected light, but when viewed in transmitted light it is decidedly ruby red. Analysis has shown it contains nanoparticles of gold. The Romans, it seems, knew about purple of Cassius long before Cassius's experiments.

Why is the Statue of Liberty green?

The statue is covered with a thin patina composed of copper carbonate and copper sulphate, both of which are green. The Statue of Liberty, properly called "Liberty Enlightening the World," sits on Liberty Island in New York Harbor. It was designed by Frédéric-Auguste Bartholdi and shipped in 350 crates as a gift to the United States by France. Liberty is fifty metres high and is covered with copper plate with an average thickness of two millimetres. More than eighty tonnes of copper were used to complete the statue. Copper reacts readily with carbon dioxide in the air, as well as with sulphuric acid produced when sulphur-containing fuel is burned. The torch Lady Liberty proudly holds aloft is covered with a thin layer of pure gold. Not much gold was needed to do this, since the metal is extremely malleable. In fact, a piece of gold the size of a matchbox can be hammered into a thin sheet large enough to cover a tennis court.

What was the first natural dye substance to be duplicated synthetically?

Alizarin, the red colourant that is found in the root of the madder plant. Before the advent of synthetic colours, dyes had to be derived from natural sources such as molluscs, insects, minerals and plants. The madder plant was grown as early as 1500 BC by the ancient Egyptians, and its ground-up root was used to dye fabrics. King Tutankhamun's tomb contained cloth dyed with madder. The red pigment isolated from the root was also known as Turkey red after the country that supplied much of the pigment throughout history.

By the Middle Ages, madder was also being cultivated in Europe and was one of the most common dyes used for fabrics, including the famed red coats of the British army. In order to bind the dye to fabric more strongly, it was usually combined with alum (aluminum sulphate), which acts as a mordant. This term derives from the Latin word for "to bite" because it allows the dye to bite into the fabric. Actually, the alum converts the water-soluble dye into an insoluble pigment that doesn't wash out. The traditional Suffolk pink that was commonly used to dye English cottages was also formulated with madder.

The plant was grown widely in Britain, but madder farms disappeared almost overnight when, in 1868, German chemists Carl Gräbe and Carl Liebermann synthesized alizarin from anthracene, a chemical isolated from coal. The chemical synthesis produced alizarin far more cheaply than the plant. Interestingly, William Henry Perkin, who in 1856 had made the first-ever synthetic dye, mauve, also came up with a synthetic method for alizarin but filed his patent one day after Gräbe and Liebermann. Alizarin became a key product for the nascent synthetic dye companies, which eventually grew into today's giant chemical enterprises. Gräbe and Lieberman had worked for BASF, a company that is still with us and is one of the largest chemical and pharmaceutical companies in the world.

What major change occurred in the production of M&M's candies in 1976?

Red M&M's were eliminated because of a health scare concerning Red Dye #2, which at the time was the most common red food dye in use. This dye had never been used in M&M's, but Mars Inc. decided to withdraw the red candies "to avoid consumer confusion and concern." It isn't clear exactly what confusion Mars was worried about, since the Food and Drug Administration had banned Red Dye #2 in January 1976. Maybe the concern was that, had red M&M's stayed on the market, the company might be suspected of using an illegal dye.

The evidence upon which Red Dye #2 was banned came from a couple of small, poorly carried-out Soviet studies in the early

1970s that suggested the dye caused thyroid tumours in male rats and stillbirths and deformities in female rats. These were followed by some flawed American studies, which even if correct implied that a human would have to drink 7,500 cans of coloured soda a day to reach the levels of dye that had been given the rats. Canada was unconvinced by the American studies and never banned Red Dye #2. Various rumours began to circulate about why the red dye was actually banned, with the most popular one suggesting that it was really an unapproved aphrodisiac. Cleverly, Mars Inc. never addressed this issue, anticipating the eventual return of the red candies. This actually happened in 1988, after the furor about the toxicity of Red Dye #2 had died away. With great fanfare the red M&M's were reintroduced, with some ingenious advertising hinting at the supposed aphrodisiac properties.

What happens if you pour red cabbage juice on the white part of a fried egg?

The egg white turns green. Normally, "red" cabbage juice is actually purple, owing to the presence of anthocyanins. These water-soluble compounds are responsible for the colour of many fruits and vegetables and are natural acid-base indicators, meaning that their colour varies depending on whether they are found in an acidic or alkaline environment. In the case of cabbage, the purple juice becomes bright red in an acid solution and greenish blue in a base. Just add some baking soda to red cabbage juice and watch it turn blue!

Anthocyanins are produced by plants to protect themselves against bright light. Plants need light for photosynthesis, but light can also generate free radicals in the plant's tissues that can impair the photosynthetic process. Anthocyanins fall into the general category of compounds known as polyphenols, which are antioxidants. In other words, they inactivate free radicals. We humans make good use of plants' attempts to protect themselves from excessive light. When we eat plant products, we ingest the anthocyanins, which can act as antioxidants in our bodies as well. And red cabbage is higher in anthocyanins than the green variety.

Perhaps more importantly, you can have some fun with this knowledge. Just boil some shredded red cabbage in a minimal amount of water, strain the juice, and pour some on your fried eggs. Now you can have green eggs with your ham. Of course, you'll have to cope with the smell of boiled cabbage. Worth doing; after all, how many chemical experiments can you perform where you can eat the product of the reaction?

The new Wembley Stadium in London has a restaurant commemorating which scientific discovery?

The first synthetic dye made by William Henry Perkin in 1856. The restaurant's décor features the dye's colour, mauve, and its windows face a canal that would turn a different colour depending on the sort of dye that was being made at Perkin's factory.

The accidental discovery of mauve was a pivotal moment in the development of chemistry. Young eighteen-year-old Perkin was actually trying to synthesize quinine from chemicals in coal tar to protect British troops from malaria when he noted that one of his reactions produced a colour that heretofore had been available only by extracting molluscs found on Mediterranean beaches. He recognized the importance of this discovery, borrowed money from his father, and built the world's first dye manufacturing plant. The consequences were widespread: if dyes could be made synthetically, so could other substances, like drugs. Indeed, the pharmaceutical industry has its roots in the discovery of mauve. So does microbiology. Synthetic dyes were used to stain microscope slides to highlight different parts of cells.

In the 1980s, Mikhail Gorbachev launched a crusade against the "green dragon." What was it, and why was it green?

The Soviet leader was concerned about his people's drinking habits and wanted to curtail the consumption of vodka. Much of the vodka Soviets drank was made in illegal home distilleries. The pipes in these stills was made from copper, which eventually turns green because of reaction with carbon dioxide in the air. Copper carbonate is green, as is readily evident on copper roofs. So the "green dragon" was vodka, and the green part came from the colour of the copper piping in home stills.

The colour of the eggs a chicken lays can be predicted by examining what part of its anatomy?

The earlobes. Chickens have ears, hidden beneath their feathers. There is no outer ear such as we have, but chickens do have earlobes.

The colour of the lobe varies with the breed of the chicken, ranging from white to almost black. Chickens with white earlobes lay white eggs exclusively, while birds with dark lobes lay brown eggs. The fascinating Araucana breed of chicken can even have earlobes that are a pale green or blue colour. Sure enough, they lay eggs of the corresponding hue. It appears that the same gene that determines the colour of the earlobe also determines the colour of the eggshell.

That colour is due to the presence of hemoglobin-breakdown products called porphyrins. Hemoglobin is the oxygen-transporting molecule found in red blood cells. These cells constantly break down, and new ones form. During breakdown, hemoglobin is metabolized into porphyrins, which can have different colours. The specific way that hemoglobin is metabolized into porphyrins is genetically controlled, meaning that the colour of eggs, which is where porphyrins are eventually deposited, is also under genetic control.

Hard-boiled eggs sometimes show a grey-green discoloration around the yolk. What two substances have reacted to form this colour?

Iron and hydrogen sulphide. The grey-green colour is due to iron sulphide, formed when iron reacts with hydrogen sulphide. Usually this happens around the yolk, but it can sometimes be seen on the surface of a peeled, boiled egg that has been refrigerated. Hydrogen sulphide is the classic odour of rotten eggs and forms when proteins in the egg white break down during cooking.

Small amounts of iron are always present naturally in eggs, mostly in the yolk. As the hydrogen sulphide formed in the egg white meets up with the iron in the yolk, greyish iron sulphide is produced. But if a boiled egg is stored in the fridge, the reaction continues, and as hydrogen sulphide comes to the surface it reacts with iron there as well.

The iron content of the white is less than that of the yolk, so the discoloration is usually not as noticeable unless the egg has been cooked in water that has a high iron content. Some of this iron will then have been absorbed through the eggshell, and the colour can be more intense. There is no health concern here: iron sulphide is a safe substance. To reduce the risk of discoloration, place the eggs in a pan, cover with water, bring to a boil and remove from the heat. Let stand for fifteen minutes, then plunge into cold water. Hydrogen sulphide will be drawn toward the cold surface, away from the yolk.

UNUSUAL
SCIENCE

Why did the father of dynamite regularly swallow nitroglycerin?

I'll come to the answer in a minute. But first, some historical background. In 1875, Dr. William Murrell placed a drop of liquid on his tongue while holding a mirror that reflected sunlight into a dark corner of the room. He noted almost immediately that the reflection moved back and forth as a result of vigorous pulsations of his hand. The liquid was nitroglycerin, and Murrell discovered that it could be used to treat angina.

Nitroglycerin had been first made by the Italian chemist Ascanio Sobrero in 1847 by treating glycerin with a mixture of sulphuric and nitric acids. He discovered that the oily liquid was a powerful explosive and also noted its sweet taste and tendency to give him a headache. That was enough to stimulate physicians' interest, and it soon became clear that nitroglycerin produced restlessness, rapid respiration and loss of reflexes when injected into animals. Obviously, nitroglycerin had an effect on the nervous system. Could it possibly be used as a drug?

In 1860, Dr. A.G. Field decided to find out and, as was common practice in those days, tried the substance on himself to note any effect. And there certainly was one! He recorded a throbbing sensation in his neck and a headache accompanied by a loud, rushing noise in his ears like steam passing out of a teakettle. The drug certainly had activity, but what could it be used for? There was only one way to find out: experiment! A lady showed up in Field's office complaining of a toothache. The doctor decided to try a drop of nitroglycerin. It quickly produced a pulsation in the neck, throbbing in the temples and nausea. But the toothache subsided. Field went on to use nitroglycerin for other neurological problems and got satisfactory results, although side effects were always severe. Others also experimented with nitroglycerin and reported varying results.

Then, in 1875, William Murrell decided to investigate the issue more systematically by taking different doses himself. He got the usual headaches and the throbbing pulse, which he quantified with the mirror experiment. These side effects were very similar to the ones that had been reported for amyl nitrite, a drug that had been introduced a few years earlier for the treatment of angina pectoris. Murrell then decided to try nitroglycerin for angina and found that it worked better than amyl nitrite. And physicians have been prescribing nitro for their angina patients ever since.

Interestingly, one of the early patients who benefited from nitroglycerin was Alfred Nobel, the inventor who had amassed a fortune by harnessing the explosive energy of nitroglycerin. Nobel had discovered that mixing the thick liquid with a type of clay yielded a solid material he called dynamite, which was far safer to handle than liquid nitro. Later in life, Nobel developed angina, and the only way he was able to stay at his desk and write his will, in which he spelled out the provisions for the Nobel Prizes, was by using a prescription of nitroglycerin. "It sounds like an irony of fate that I should be ordered to take nitroglycerin internally. They call it Trinitrin so as not to scare the chemist and the public," he commented.

Drs. James Nolan, Thomas Stillwell and John Sands received what award for a paper published in *The Journal of Emergency Medicine* entitled "Acute Management of the Zipper-Entrapped Penis?"

The Ig Nobel Prize in Medicine. Not quite the Nobel Prize but nonetheless important in its own right. The Ig Nobels are handed out every year at a ceremony at Harvard University, attended by more than a thousand people who appreciate journeys to the forefront of science. The sole criterion for an award is to have done something that first makes people laugh, then makes them think. Management of a zipper-entrapped penis nicely conforms to this criterion—except, of course, for the subject, who is more likely to have emitted screams of pain than howls of laughter.

The three physicians refer to the case of a young boy brought to the emergency room with his foreskin lodged in the grip of his pyjama zipper. After analyzing the situation, the doctors decided that quick action was needed and concluded that a bone-cutting device had to be applied—to the zipper, not the entrapped member. They describe how the median bar of the zipper instantly yielded to the bone cutter, freeing the organ. The parents, concluding that such an experience should be limited to once in a lifetime, asked that a formal circumcision be undertaken.

As advice to other physicians confronted with such entrapment, the award winners offer the following: "Extraction by vigorous manipulation, including attempts at unzipping the skin or prying the zipper are usually unsuccessful, painful, and can lead to further injury. Our case exemplifies a simple, quick, and nearly pain-free method of freeing entrapped zippers." As long as a bone cutter is readily available. The doctors' 1990 paper makes no mention of any use of a topical anaesthetic such as lidocaine, which would probably be a good idea. So far, there have been no activist groups demonstrating against the use of zippers in men's pants, despite the documented evidence of potential injury.

In 1984, an employee set off the radiation alarms as he arrived for work at the Limerick nuclear power plant in Pennsylvania, which was not yet operational. What was the source of the radiation?

Radon gas. Stanley Watras was as confused as anyone about why he was radioactive. He asked that his home be tested for radiation, and bingo, there it was! Levels of radioactive radon gas were found to be 650 times higher than normal. The lung cancer risk at this level is roughly equivalent to smoking 135 packs of cigarettes per day.

Where was the radon coming from? It was seeping up from the ground underneath the Watras home. The original source of radon is uranium-238, which occurs naturally in many types of rock. When uranium undergoes radioactive decay, it forms radium-226, which in turn decays to radon-222 by releasing an alpha particle. Radon itself is not dangerous, but the problem is that it undergoes very quick radioactive decay with a half-life of just 3.8 days to form polonium-218, which is a solid that attaches to dust particles and ends up being inhaled.

In the lungs, polonium undergoes further decay, emitting alpha particles that can damage cells and cause lung cancer. The only way to know if a home is contaminated with radon is to use a radon detector. This is a little charcoal canister that is usually placed in the basement and adsorbs radon gas if it is present. It is then sent for analysis. The unit used to measure radiation is the picocurie, named after Marie Curie. It describes the number of atoms undergoing radioactive decay per minute. The average indoor level is under 1.5 picocuries per litre of air, and the safe limit is 4 picocuries. The Watras home had an astonishing level of 415 picocuries per litre. The house just happened to be sitting on top of granite deposits that contained a high level of uranium, and there were fissures in the earth where the gas was seeping up. When high levels of radon are detected, a ventilation system can be installed beneath the house to vent the gas into the outside air. Cracks in the foundation have to be sealed as well. Radon is colourless and odourless and unfortunately is often ignored. It is the second-leading cause of lung cancer after smoking.

Granite countertops, contrary to alarmist silliness on the Internet, are not a significant source of radon gas. Yes, there may be traces of uranium in the granite, but the amount is inconsequential. A granite countertop may make a dent in your wallet but not in your health.

Thomas Becket's body lay in Canterbury Cathedral overnight after his murder. It was written that, "As the body grew cold, the vermin that were living in this multiple covering started to crawl out and boiled over like water in a simmering cauldron." What vermin?

Body lice. Sounds absolutely repugnant, but body lice have plagued humans throughout history. Frequent bathing and wearing clean clothes, the best weapons against lice, are relatively recent developments. Lice frolic in the warm, moist environment produced by wearing layers of clothing. Becket, the Archbishop of Canterbury, perhaps not your typical example, wore a large brown mantle, a white surplice, a lamb's wool coat, a black cowled robe and a curious haircloth covered with linen. One wonders how he moved. But the lice inside his clothing certainly did—and, when possible, hopped onto others.

Infestation with body lice was common, causing people to itch and scratch the bites. This is where the real problem started. A louse that had feasted on a person infected with the bacterium that causes typhus (*Rickettsia prowazekii*) was capable of transferring the microbe to another person, who by scratching a bite allowed louse fecal matter that harboured the bacterium to be transferred into his bloodstream.

The closer the contact between people, the greater the likelihood of a typhus epidemic. And there have been plenty of those. War, with soldiers and prisoners living under crowded, dirty, brutal conditions, is ideal for the spread of the typhus bacterium. The results are not pleasant. Typhus, a completely different condition from typhoid fever, produces a rash, muscle pains, cough and a high fever that can kill. In the eighteenth century, being imprisoned in Britain was tantamount to a death sentence from typhus. Napoleon's army suffered more casualties from the disease than from Russian weapons, and millions died of typhus during the First World War.

Typhus epidemics blazed through Nazi concentration camps, and German soldiers invading Russia suffered the same fate as Napoleon's army. Allied forces were not as widely affected thanks to dusting with DDT, an insecticide that is very effective against lice. Refugees and displaced people were also treated with DDT spray, preventing potentially horrific epidemics. A vaccine against typhus had been developed by 1943, greatly reducing the risk of future outbreaks.

In 1935, Harvard microbiologist Hans Zinsser chronicled the story of typhus in his epic work *Rats, Lice and History*, drawing attention to the importance of fighting microbes. Not only did it provide a fascinating account of how history was shaped by microbes, it had a huge impact, in a roundabout way, on conquering measles. John Franklin Enders was a young millionaire who whiled away his time studying English at Harvard when he chanced upon Zinsser's book. He was so taken by it that he switched to microbiology, earned a PhD and found a way to grow viruses to produce safe vaccines. He received a Nobel Prize for this work and went on to develop a vaccine against measles that has saved millions of lives. All because he read about body lice.

Why would forensic investigators apply a solution of copper chloride to a gun?

To read serial numbers that had been filed off. Serial numbers are stamped into the metal with a hardened steel die, compressing the metal underneath, squeezing the metal atoms closer together. Iron is an active metal, and it undergoes a variety of chemical reactions, the most famous of which is corrosion. In this process, iron atoms give up electrons and become iron ions, transforming the surface from a hard material characterized by an ordered arrangement of metal atoms to a crumbly ionic compound. This reaction can happen quickly in the presence of copper chloride. Essentially, electrons are transferred from the iron to the copper. As the iron is oxidized to iron oxide, or rust, the copper ions are reduced to metallic copper. This process occurs more quickly in the region of the metal where the iron atoms have been compressed together, so that the iron dissolves away faster than in the surrounding area and the serial numbers appear. When the numbers appear, the reagent is neutralized with a base and the surface is photographed.

What insect is named after an old wives' tale?

The earwig. They're ugly, they can "bite" with their pincer-like appendages, they can emit a nasty smell and they can damage flowers and vegetables. But they do not climb into the ears of sleeping people and bore into their brain. It's hard to know where this silly notion originates, but it may be based on the earwig's preference for hiding in dark, humid spaces. They're nocturnal creatures and take shelter during daylight hours under rocks, in bundled newspapers or crevices in kitchen shelving. Some are up to date on modern technology and seek refuge in computer keyboards. There is no evidence, though, that earwigs seek out ears.

They're voracious eaters, happily dining on insects, foliage, fruits and grains. In general, they do not present a great risk to crops or to home gardens, but their appearance is unnerving to some people. Controlling a heavy infestation of earwigs requires synthetic pesticides such as diazinon, propoxur or bifenthrin, while lighter infestations can be addressed with a plant oil–based product called EcoDust that is harmless to humans.

Before resorting to pesticides, it is best to try to cut down on earwig population by using traps. These can be very simple to make. A cardboard box with oatmeal for bait and holes punched near the bottom with a pencil works well. So does a rolled-up newspaper or the centre roll from a paper towel. Just dump these, with their content of earwigs, into a bucket of soapy water and that will take care of the pests. But the best may be an amber-coloured beer bottle with a couple of inches of flat beer in it. Bury the bottle in the garden with about an inch sticking out. Earwigs apparently are partial to beer and will be attracted by the smell, climb into the bottle and drown. Do not drink the beer after!

While earwigs crawling into an ear is the stuff of urban legends, it does make for a good story. At least Rod Serling, the creator of the 1970s TV show *Night Gallery*, thought so. In a thrilling episode confusingly called "The Caterpillar," a man falls in love with a young woman who happens to be married. He figures that his chances for her affection will improve dramatically if he can get the husband out of the way. On hearing a story about earwigs crawling into the brain, our hero decides that this may be the way of getting rid of his competition without suspicion falling on him. So he hires a drunken sailor to implant an earwig in the man's ear. But drunken sailors are not great at following instructions, and the sailor ends up putting the earwig into the wrong ear—namely, that of the would-be murderer. The earwig bores into his brain, causing excruciating pain, but eventually emerges from the opposite ear without killing the victim. The episode closes with the victim, who miraculously survived this earwig attack, finding out that the insect was a female and laid eggs in his brain. Justice was served.

While you don't have to worry about an earwig eating your brain, molesting one may leave you with an unpleasant-smelling brown stain on the fingers. Earwigs have a fascinating defence mechanism. When perturbed, they discharge a spray from a pair of defensive glands on their abdomen. The toxic and stinky chemical mixture of quinones and pentadecane can effectively deter predators.

Why was Alexander Graham Bell called to U.S. President James A. Garfield's bedside after he was shot by a would-be assassin?

To try to locate the bullet that had entered the president's body. Bell is best known for his invention of the telephone, but he also invented the metal detector. A primitive form of this device was quickly put together when Garfield's doctors could not locate the bullet that they feared would kill the president.

Garfield had been shot by Charles Guiteau, a madman who believed that God had ordered him to carry out the assassination. Guiteau fired two shots at the president, one of which grazed his arm while the other ended up somewhere in the abdomen. Garfield was immediately taken back to the White House, where doctors poked his wound with unsterile fingers trying to find the bullet. When they were unable to locate it, government officials summoned Bell, who had already suggested that he might have a way of finding the bullet based on the principles he had used to make the first telephone.

Bell's telephone was based on the idea that a current is generated in a wire as it moves through a magnetic field. Similarly, the metal-detecting device he designed consisted of a coil of wire that generated a magnetic field when an alternating current passed through it. This magnetic field in turn induced another one in a nearby metallic object, and *that* field was detected by its effect on the electric current passing through a second coiled wire. This was the gadget that Bell brought to the White House in 1881 after sixteen physicians had failed to locate the bullet in the president's body.

The physicians were so skeptical of Bell's effort that they refused to move the president to another bed when Bell complained that the metal frame and coils were interfering with his metal detector. In any case, Bell claimed the bullet was located deep in the body. An autopsy revealed that the bullet had ended up about four inches from the spine and had not done great damage—but the doctors had! Eighty days of probing the wound with dirty fingers and forceps had caused an infection that precipitated the president's death.

The treatment the poor man received was brutal, but was par for the course at the time. The physicians even submitted a bill for

$85,000 to the government for services rendered, but they received only $10,000. Bell went on to great fame, but he did not pursue development of the metal detector any further. Credit for a workable device goes to Gerhard Fischer, who in 1925, while working with a portable radio apparatus, found that he was getting interference when he walked past a large metal tank. His metal detector went on sale in 1931. Little did he imagine that it would be used in some form all over the world to fight terrorism and crime.

A sample of hair was sold on eBay for more than $200,000. What was unusual about it?

It was Beethoven's, and it had been turned into a 0.56-carat diamond. The gem was created by an American company specializing in making diamonds from unusual sources of carbon, such as the remains of human and pet bodies. (The $200,000 was donated to aid underprivileged children.)

Diamonds are actually crystals of carbon in which each atom is bonded to four others in a three-dimensional network, a pattern that makes the diamond the hardest known material on earth. Very high temperatures and pressures are required to fuse carbon atoms into this particular arrangement, conditions that were met millions of years ago in molten lava, miles below the surface of the earth. Volcanic eruptions subsequently brought the diamonds to the surface. Today, the conditions needed to produce diamonds can be duplicated in the laboratory. Artificial diamonds have been produced since 1954 and are essentially indistinguishable from the natural variety. Any source of carbon will do, including organic compounds found in the human body.

LifeGem, which created the Beethoven gem, was the first company to capitalize on that idea and offer "memorial diamonds" that ensured that loved ones could be kept close to the heart in more than just spirit. After cremation, the remains are converted to graphite, which is then subjected to high pressure and temperatures in the range of 1,600°C to 2,000°C for six to nine months, depending on the colour of diamond desired. A single body contains enough carbon to make about fifty one-carat stones, but generally a customer only wants one. Prices reach $14,000 per stone. Although the business may seem a little morbid, it is booming, with new companies jumping on the bandwagon.

What flavouring agent can be derived from cow dung?

Vanillin, the major flavour compound in vanilla. The most desirable vanilla flavouring is extracted from the seed pods of *Vanilla planifola*, an orchid that was originally native to Mexico but now is grown mostly in Madagascar. Hernán Cortés was probably the first European to learn about vanilla, back in the sixteenth century, when the Aztecs treated him to *chocolatl*, a beverage made from cocoa beans and flavoured with vanilla pods. For about 350 years, if you wanted to produce vanilla flavouring—and plenty of people did—you'd better know where to find a vanilla orchid. Extracting the pods with alcohol yielded a complex mixture of compounds with a delicate vanilla flavour. When chemists got into the game, they discovered that most of the flavour was created by one compound, vanillin, which could also be synthesized in the laboratory.

By the late 1800s chemists were able to make vanilla on a large scale from eugenol, a compound found in the oil of cloves. As demand for the flavouring agent grew, novel and more efficient methods of production were sought. It turned out that lignin, a complex molecule that forms part of the cell wall of plants, could be efficiently converted into vanillin, and the industry switched to producing vanillin from pulp and paper waste. And when chemists discovered that vanillin could be readily made from guaiacol, a compound obtainable from petroleum, the current method of production was born.

But now, with the green movement in full swing, substances not derived from petroleum are highly marketable—especially if they can be made from some sort of waste material. Like cow dung. Mayu Yamamoto at the International Medical Centre of Japan found that grass-eating animals excrete a lot of lignin in their waste, and this can serve as a source of vanillin. The raw material is readily available, and the extraction process is simple. Cow dung–derived vanillin can be produced for about half the price of the extract derived from vanilla beans. A cow dung–processing machine is presently being designed for just this purpose.

There is no intention to market cow dung vanillin as a food flavouring, although the compound is identical to that produced by any other process. The concern is that people may be wary of this vanillin if they learn of its origin. So the cow dung vanillin will be used as a fragrance in perfumes, hair care products and skin care formulations. It seems that the cosmetic industry is not as worried about the cow dung association as the food industry is. After all, they already use ambergris, which is whale vomit, as a starting material to make fragrances.

IT *IS*
ROCKET
SCIENCE

Most satellites are launched in an easterly direction. Why?

To take advantage of the speed of the earth's rotation. In order to place any object into orbit around the earth, be it a satellite or the space shuttle, two conditions have to be met. A height of at least 160 kilometres has to be achieved to ensure that the atmosphere is thin enough not to cause a slowdown due to friction with air. The other requirement is a speed of about 28,000 kilometres per hour, equivalent to an astounding 8 kilometres per second, in order to cancel out the pull of gravity. That's much faster than a speeding bullet!

Any object that has been launched and meets these requirements will go into orbit around the earth. Once orbital height and velocity have been achieved, there is no longer any need for power. Since there is no air to retard forward motion, the object will orbit forever, at least in theory. In other words, it will be in a continuous state of free fall around the earth. In reality, even at that height there are traces of air, so that eventually there will be enough of a slowdown due to friction to allow gravity to pull the orbiting object back to earth.

Now, why the easterly launch? Because the earth rotates in an easterly direction. The speed of rotation varies according to latitude; it's at its highest at the equator (about 1,600 km/h), its lowest at the poles. At Cape Canaveral the speed is about 1,500 km/h. This means that to achieve orbital speed, a launch vehicle only has to accelerate from 1,500 km/h to 28,000 km/h. If it were launched westward, there would be a speed penalty of 1,500 km/h, meaning that the launch vehicle would have to accelerate to 29,500 km/h to achieve orbital velocity. The greater the speed that has to be achieved, the greater the technological and fuel requirements.

Placing a satellite into an orbit that circles the poles requires a slightly westerly launch. Such launches usually take place from California instead of Florida, so that if something goes wrong, debris will fall into the ocean rather than over populated areas. Israeli satellite launches also take place in a westerly direction to avoid a launch path over unfriendly neighbours. Shuttle landings are also in an easterly direction to take advantage of the earth's rotation. Since the earth is moving in the same direction as the landing shuttle, it doesn't have to slow down as much before touching the ground as it would if it were landing in a westerly direction. This is why, when the *Columbia* tragically disintegrated on re-entry into the atmosphere, debris was strewn mostly over Texas. The shuttle was entering the atmosphere over California, heading toward Florida.

What is the use of potassium superoxide, K_2O, in submarines and space vehicles?

It can be used to absorb carbon dioxide and at the same time generate oxygen. The oxygen required for breathing aboard submarines and space vehicles is stored in oxygen cylinders. The exhaled carbon dioxide is commonly absorbed by a chemical solution of sodium or calcium hydroxide. But there always has to be a backup system in case the oxygen runs out. This is where potassium superoxide, which is made by burning potassium metal in oxygen, comes in. It serves a double purpose. When it reacts with carbon dioxide it forms potassium carbonate, and at the same time releases oxygen gas. The Russians have even used this chemistry to supply the oxygen needed during spacewalks.

The lunar lander's descent to the moon relied on an engine that used no ignition system. The fuel ignited when it came into contact with the oxidizing agent. What do we call such a system?

Hypergolic. The giant three-stage Saturn V booster that propelled the astronauts toward the moon in 1969 was the most powerful rocket ever built. All stages used liquid oxygen as the oxidizing agent, but the fuels were different. The first stage used a variety of kerosene, while the second and third stages used liquid hydrogen. The engines used in all stages were very sophisticated, but most people don't realize that probably the most technologically advanced equipment of the lunar missions was the engine used to take the lunar module to the moon's surface and later to lift it off again.

To land safely on the moon, the module needed to be slowed as it came out of lunar orbit and hurtled toward the moon's surface. Parachutes would be useless, since the moon has no atmosphere. The only way to slow the descent was by firing an engine in the opposite direction—that is, toward the surface. The hot gases jetting out then slowed the vehicle according to Newton's third law: For every action, there is an equal and opposite reaction.

All rocket engines rely on producing the exhaust gases by burning a fuel in the presence of oxygen. The oxygen is either stored as pressurized liquid oxygen or released as needed by a chemical referred to as an oxidizing agent. But just as starting a fire needs some sort of ignition, so does a rocket engine. Engineers feared using an engine that required ignition for the moon landing because failure would have been catastrophic. They also required an engine that could be controlled with a throttle in case the astronauts had to modify the descent and hover over the lunar surface should a change in the landing site be necessary. That is why the decision was made to go with a hypergolic system.

The oxidizing agent was dinitrogen tetroxide, and the fuel was a fifty-fifty mix of hydrazine and unsymmetrical dimethylhydrazine. When these are combined, ignition is instantaneous. But there was another problem that had to be addressed: how do you move the propellants from the storage tanks into the engine? In the weightlessness of space, the contents of the tanks just float around. Clever engineering came to the fore here as well. The oxidizer and fuel were stored inside a bladder made of Teflon inside the tanks. When ignition was needed, pressurized helium was introduced by turning a valve, creating pressure on the bladder, which pushed the propellants into the lines to the engine. It turned out that this design paid dividends.

As the astronauts approached the moon's surface, the commander, Neil Armstrong, realized that they were coming down toward the side of a crater. Had they continued on this path, the module would have turned over, with a tragic ending. Armstrong took over the controls and executed a hovering manoeuvre by controlling the valves that released the pressurized helium. And then we heard the classic words: "Houston, Tranquility Base here. The

Eagle has landed." If it had not been for the manoeuvrability of the hypergolic engine, the landing would not have been so tranquil.

The same engine was used to successfully lift the module from the moon's surface. You could not take a chance on pushing an ignition button and hearing *grrrrrrrrrrrrrrrrrr*. There would have been no rescue possible.

The first synthetic rubber manufactured in the U.S. was named after the Greek words for sulphur and glue. What was it called?

Thiokol. In 1927, Kansas City chemist J.C. Patrick was looking for a cheaper and better alternative to ethylene glycol antifreeze. In the process, he reacted dichloroethane with sodium polysulphide and, instead of a liquidy antifreeze, ended up with a solid rubbery substance. He had accidentally created a polymer composed of alternating sets of carbon and sulphur atoms strung together. Since the material was sticky and incorporated sulphur, he named it Thiokol, from the Greek *thio* for sulphur and *kol* for glue. He even started a company to produce the product and named that Thiokol as well.

The rubbery substance was easily recognizable because of its foul smell and was quickly dubbed "synthetic halitosis" by people who worked with it. Although never a replacement for rubber, Thiokol found use as a sealant for aircraft carrier decks and in parts of hoses exposed to petroleum products, to which it proved highly stable. Thiokol's rise to fame, however, did not stem from its sealant properties. In the 1940s, Caltech's Jet Propulsion Laboratory was working on developing solid-fuel rocket engines to replace liquid-fuelled engines that were hard to handle and required a maze of tanks, valves and pumps. Solid-fuel rockets require an oxidizing agent and a fuel to be blended together in a uniform fashion, and Thiokol turned out to have just the right "gluing" properties to do that. Better yet, it was itself flammable and therefore also acted as a fuel on top of being an excellent binder. Indeed, the solid-fuel boosters of the space shuttle still use this technology today. And it all came from a search for a better antifreeze.

INDEX